Modern Census to Identify the Deprived and Vulnerable

Modern Census to Identify the Deprived and Vulnerable

Technology and Institutions in Managing SECC

Editor

N.K. Sahu

ACADEMIC FOUNDATION
NEW DELHI

www.academicfoundation.org

First published in 2018
by

ACADEMIC FOUNDATION
4772-73 / 23 Bharat Ram Road, (23 Ansari Road),
Darya Ganj, New Delhi - 110 002 (India).
Phones : 23245001 / 02.
Fax : +91-11-23245005.
E-mail : booksaf@gmail.com
www.academicfoundation.org

Disclaimer:
The findings/views/opinions expressed in this book are solely those of the authors and do not necessarily reflect the views of the publisher.

Cataloging in Publication Data--DK
Courtesy: D.K. Agencies (P) Ltd. <docinfo@dkagencies.com>

Modern census to Identify the deprived and vulnerable : technology and institutions in managing SECC / editor, N.K. Sahu.
pages cm
Contributed articles.

1. Socio-Economic and Caste Census Project (India). 2. Caste--India--Statistics. 3. India--Census. 4. India--Census--Data processing. 5. Project management--India. I. Sahu, N. K. (Naba Krushna), 1958- editor.

LCC HA4585.M63 2018 | DDC 315.4 23

Typeset, printed and bound by The Book Mint, New Delhi.
www.thebookmint.in

CONTENTS

Part 2
Project Management

Part 3
Technology Support System

LIST OF TABLES AND FIGURES

TABLES

FIGURES

About the Editor/Contributors

Editor:

N.K. Sahu belongs to Indian Economic Service (IES). He was the Economic Adviser in the Ministry of Rural Development (MoRD), Government of India, the nodal minstry, which carried-out the Socio-Economic and Caste Census (SECC). He is the only person who worked for this project since its inception till it came to the final stage. His role was unique as he worked for the predecessor of this project, i.e. BPL Census 2002 as well as the SECC project. His PhD thesis is on "Analysis of Forced Commerce in Indian Agriculture" (unpublished) and has authored a book titled: *Electoral Politics in Federal India: MP Local Area Development Scheme*. At present, he is the Economic Adviser, Ministry of Human Resource Development, Government of India, New Delhi.

Contributors:

Joseph Abraham was with the Kerala State Service (KSS) for more than two decades as a development expert and possesses work experience in rural/urban/IT and tourism sectors. He had worked with National Institute of Rural Development (NIRD), Hyderabad and Jaipur as a consultant cum faculty in rural economics. For the last four and half years, he worked as Social Scientist with the MoRD in the SECC unit in New Delhi. Besides PhD, he has 2 books published, 25 peer reviewed publications and about a dozen reports on rural development in India. Currently, he is engaged in promoting eco-friendly framing and agrarian progress besides keeping abreast with the developments in poverty reduction programmes and schemes like DDU&GKY, MGNREGS, PMAY and PMJDY etc. in tandem with the findings of SECC 2011. He was the Principal Social Scientist of SECC, MoRD, and presently is Senior Fellow, Indian Council of Social Science Research (ICSSR).

Purnendu Kishore Banerjee belongs to Indian Statistical Service (ISS). During SECC project executions, he worked as the Deputy Registrar General in the Office of the Registrar General and Census Commissioner, India. He has worked in both the Population Census and the SECC projects. Prior to that, he worked in the National Sample Survey Organisation. At present, he is working in the National Accounts Division of the Central Statistics Office, Ministry of Statistics and Programme Implementation, Government of India.

Partha Pratim Bhattacharjee worked in SECC project from start to end as a technical expert. He was a technical consultant in the project. Before working in this project, he worked with Indian Railway (Signal & Telecomm), Indian Institute of Technology (Kharagpur) and National Institute of Rural Development and Panchayati Raj (NIRD&PR), Hyderabad. Presently he is Assistant Professor, NIRD&PR, Hyderabad.

Savita Kulkarni is an Independent Gender Consultant. From September 2010 to March 2013, she was Consultant with UN Women India and was deputed to MoRD, Government of India, for SECC 2011. She has experience of working in Afghanistan, Mozambique and Myanmar.

Tej Tarun Mahindra is working as Manager at Bharat Electronics Limited (BEL). He worked as nodal officer from BEL for the SECC project and was deputed at MoRD from September 2011 to March 2016. He was involved in software development for various modules used in SECC along with coordination among CPSUs (central public sector undertakings), MoRD and state officials. He was also involved in various training programmes for the SECC. He did his Masters in Engineering from Delhi Technical University (formerly known as Delhi College of Engineering).

Varindra Seth is working as Technical Director at National Informatics Centre, Ministry of Electronics & Information Technology, Government of India. He worked as head of SECC project for developing the central MIS (management information system), its hosting and maintenance. He did his Masters in Technology from Indian Institute of Technology, Roorkee (formerly known as University of Roorkee). Presently, he is Scientist 'E' in National Informatics Centre, New Delhi.

Pramod Sharma is working as Principal System Analyst at National Informatics Centre, Ministry of Electronics & Information Technology, Government of India. He assisted SECC project in developing the central MIS, its hosting and maintenance. He did his PhD in Microelectronics from Agra University, Agra. Presently, he is Scientist 'D' in National Informatics Centre, New Delhi.

Santhi Sozhan is working as Senior DGM in Bharat Electronics Limited, Bangalore. She had worked in the design, development, production and customer support of the TPCs (tablet personal computer) which were used in the SECC project. She has knowledge of embedded technology products and processes.

Sudhir Kumar Tewari belongs to Indian Economic Service (IES). During SECC project executions, he worked as the Economic Adviser in the Ministry of Housing and Urban Poverty Alleviation (MoHUPA). He has held senior level posts in the Ministry of Finance, Ministry of Agriculture and Cooperation, and worked in the Finance Department of the Government of Madhya Pradesh. He has contributed on many items in Arthapedia—a portal for facilitating understanding of the Indian economy and its governance by explaining the concepts used in the economic policy domain in India (*http://www.arthapedia.in/*).

Preface

Conducting census is normal in most of the countries. But in India, besides decennial census of population, census of rural households was conducted at regular intervals to identify government scheme beneficiaries. Though it was conducted in India in 1992, 1997 and 2002 under the name 'BPL Census' to identify rural households that are below the poverty line, it is doubtful that we may find the data as well as the documents that captured the processes followed to organise these census. Despite the use of information technology and the fact that computers have been in vogue for long, including in these BPL Census, even the electronic footprints of such exercises were not easy to find. Thus, it became imperative on the part of the team that worked for the SECC (socio-economic and caste census) project to record the procedures followed and management principles adopted in a book (hard copy) format. This would help the posterity to gain in various ways and also prevent lost of such a socio-political project in bewilder. I personally feel, such projects come in the life of a big country like India, very rarely. This is a project that collected socioeconomic and caste data of all the households of a vast and diverse country in a transparent manner with peoples' participation. This is about management of big project of socio-political nature. Of course, taking advantage of information and computer technology is another aspect of this project. This book would be a useful reading for students and practitioners of management, statistics, demography, public policy and administration, sociology, and social sciences.

The SECC project preparation started long before 2011 and the real work was carried out in the 1st half of the present decade. The experience and insights we gathered in course of implementing this project is really unique and valuable. Probably the world expects us to express, speak or record various aspects of planning, development and execution of SECC. Accordingly this book is compiled covering key features and challenges of the project. Thus, this book has been written by a handful of SECC project key executioners capturing the contributions made by thousands of individuals and institutions. On behalf of these authors, I acknowledge the contributions of each and all. The officials of various ministries in the government of India, state governments and public sector units deserve thankful appreciations for permitting and encouraging all of us to publish this book.

—N.K. Sahu

N.K. SAHU

Introduction

Socio-Economic and Caste Census (SECC) project aimed at gathering household and individual level data throughout India, both in rural and urban areas in census mode. This project transcended the normal statistical process of data collection as is done world over in either census or sampling modes. Conventionally the data collected through these processes does not display the individual/household characteristics of the respondents and primary units. Such data are subject to analysis and use only at aggregate level. One can get village level aggregate data collected through decennial population census in India. Similarly, these days, the unit-level data are made available for National Sample Surveys being collected by National Sample Survey Office (NSSO), but, without the identification particulars of the respondent units. However, as the objective of SECC is mainly to facilitate strict correspondence between the targeted government benefits and intended individuals and households, the relevant particulars of the survey units are put to public scrutiny while meeting their privacy and secrecy concerns. Thus, while the SECC data was subject to public scrutiny, the sub-set of respondents who wanted to maintain their privacy were kept out of its purview. Therefore, SECC had gone beyond a normal statistical exercise and encapsulated various dimensions of a complex sociopolitical process. The outcome of such an exercise with regard to rural areas was made public in July 2015 and for urban areas in November 2016. The data pertaining to 'caste' status of individuals is yet to be made public.

This project was implemented in an arena comprising interplay of men and machines within an institutional setup. About a million individuals were involved in collection, collation and analysis of demographic and socioeconomic data of more than billion people. They all worked within the sociopolitical environment and the given boundaries of a federal-democratic constitutional framework. The book explains the way the SECC project was executed. Many of the processes were novel and thus encountered uncertainty. But a bold decision was taken to conduct the nation-wide census as the expected benefit of such an exercise would be immense. Many lessons were learned while executing the project and, that was probably one of the valuable learning outcomes derived by its implementers. In this introductory chapter, the objective is to summarise the contents of the subsequent chapters. Listing out the main lessons that were learned by the stakeholders while implementing the project would be made in the concluding chapter of this book. Though the experience of the ground level enumerators is no less valuable than the team that worked at the national and state level, the aim here is to accommodate everybody's concerns and pen-down only the crucial aspects.

SECC has made valuable methodological and empirical contributions. How to identify a household and individuals on the basis of multiple deprivations and/or vulnerabilities for targeting scheme/project was the main problem faced by a government. However, once a method is finalised, the next step is to identify such households or individuals based on information gathered from various sources. In India, the government agencies mostly depend on field surveys to identify such target beneficiaries. But census methods were also resorted to in many cases. All these conceptual issues and the empirical methods that are followed are explained in the chapter 1 by Abraham. The efforts to take advantage of the existing data base from decadal Population Census 2011 as the basic building block and the readiness to use the census concepts with appropriate modifications in SECC has proved to be very valuable. The readiness to work together by the Census, Rural and Urban Nodal Departments to identify and locate the targeted families and individuals resulted in accomplishment of a complex and vast project that reinforced and rectified the strengths and weakness of each other's approaches and systems of data collection. They learned from each other and improved the quality of outcomes. Ministry of Rural Development (MoRD) believes in peoples participation and transparency to achieve accountability, and thus credibility and the experience in holding Gram Sabha (village council) meetings in its BPL (below poverty line) Census 2002 in rural areas was the most valuable background for organising the SECC project. In SECC, holding of Gram Sabhas was made compulsory for scrutiny of the Draft Lists that contain household and individual member information along with socioeconomic indicators (except religion and caste status). Due to the peculiarities of the PRIs (panchayati raj institutions) in the urban areas, it was decided not to go through the public scrutiny in the Urban Wards Sabhas. However, the public scrutiny in urban areas was achieved through two stages, viz.: (i) public display of Draft List, and (ii) claims disposal. Delicate balance between 'privacy requirements' and 'public scrutiny' could be achieved in SECC by carefully collecting the consents of only the agreeing respondents at the end of enumerations and publicly collecting dissents after proper advertisements. This was imperative as limited numbers of targeted beneficiaries are to be selected under government schemes and their selection based on socioeconomic parameters of the households and/or individuals needs public scrutiny. On the other hand, those who don't aspire for government benefit and value 'privacy' are kept outside the purview of public scrutiny.

The project gained immensely from the experience of the Registrar General and Census Commissioner of India (RG&CCI)'s methodical and meticulous approach in conducting decadal census since 1872. SECC was conducted utilising the house listing data and NPR data collected by RG&CCI as the base and updating the demographic data where ever required.

Ministry of Housing and Urban Poverty Alleviation (MoHUPA)[1] administers various schemes, projects and oversees the legal and regulatory framework that aims at improving the living conditions of millions of urban dwellers sustaining themselves in varied

1. Since merged with the then Ministry of Urban Development and the combined fraction has been renamed as Ministry of Housing and Urban Affairs.

socioeconomic environment. The SECC project learned from them the way to handle complex and densely populated settlements that exist in Indian urban ecosystems. The socioeconomic data of urban households was collected mainly to capture the residential vulnerability, social vulnerability, occupational vulnerability and housing related vulnerabilities. Care was taken while rural-urban classifications of areas differ between RG&CCI and MoHUPA in terms of Census Towns and Statutory Towns respectively. In chapter 2, Tewari and Sahu explain the challenges faced in conducting SECC in urban areas and the steps taken to address these. While, as mentioned earlier, house listing data of RG&CCI was the base of SECC field enumerations, rural-urban demarcation became one of the difficult operations. This challenge was also handled by the project team. Similarly, though the census operation was done through digital mode, converting the scanned uploaded images of NPR (national population register) in the TPCs (tablet personal computers) to digital format posed technical challenges. These two technical issues are explained in chapter 10 by Bhattacharjee.

The substantial presence of public and private corporate sectors in performing many crucial activities also throws many opportunities as well as challenges. Throughout the project cycles, the knowledge systems of all these agencies were harnessed and about a million of human resources were engaged through them to achieve the project objectives—collecting household and individual demographic and socioeconomic data in both rural and urban areas of the country. This project was implemented in a transparent and participatory mode that resulted in credibility of outcomes. The methods followed to achieve the resultant credibility are explained in chapter 4 by Abraham. Transparency, accountability and public participation was ensured at each stage of SECC process; starting with canvassing of questionnaire at enumeration stage to organising the public scrutiny protocols. India is democracy and the role given to PRIs in a federal system can be seen in full play in case of SECC. Since credibility and reliability of outcomes of SECC is paramount to government, a number of processes, as mentioned above, were specially designed to achieve these objectives. However, many a time the democratic processes are subverted due to the local fragmented politics of powerful groups within local areas. To make up for these deficiencies, it was decided to put in place a third party concurrent monitoring system. The details of such a system is given in chapter 6 by Sahu, explaining the challenges faced while implementing it and deriving the benefits of such a system.

In chapter 5, Sahu explains the role played by various stakeholders in the SECC project as success of any project depends upon the role played by each of the stakeholders. Knowing the nature of stakeholders and the stakes in such a big and complex project of sociopolitical nature is the key to managing them properly to achieve the main goals of the project. Thirteen (13) stakeholders and about 17 stakes were identified and analysed in this chapter. It shows the complex role played by the stakeholders and the communication and engagement structure that was constructed and adhered to by each of the key stakeholders. Managing such a big and complex project with multiple stakeholders working with about a million individuals, lakhs of electronic devices like TPCs, thousands of local computers/servers and high capacity central servers was possible and made effective by establishing and operationalising a very large database and MIS (management information system). These aspects of SECC are narrated in chapter 11 by Seth, Sharma and Sahu. The adoption of fully digital operations facilitated an active MIS that added enormous value to the whole exercise. Close and in-time monitoring was possible and various solutions were provided to all levels of operators from field to Central government. For example, unlike in case of BPL 2002 Census, the number and name of villages that hold Gram Sabha meetings and the numbers and type of claims and objecting petitions received and the action taken on all such petitions was closely monitored. Huge meta-data could be generated and put to use for adding value to the SECC project was possible as the whole exercise was done in digital mode using computer and communication technologies.

SECC was conducted using the latest available informatics and computer technologies in order to cover an estimated 1.2 billion population spread over a vast and diverse terrain in a time bound manner. As the outcomes of the project would help the government to identify households and individuals for targeting scheme benefits, quality of such outcomes are of paramount importance. Thus it was decided to collect data through TPCs. There were two problems in this regard. First, to procure more than 6 lakh TPCs at a time when the capacity of the country to produce and use such instruments was limited to only a few thousand. The challenges of procuring 6.4 lakh TPCs, delivering these at hundreds of locations and ensuring its smooth functioning was explained in chapter 8 by Mahindra and Santhi Sozhan. Second, customising all these TPCs not only as instruments of data collection but also to use it for updating legacy and pre-collected data was a challenge. When such data is not fully in digital format (images of NPR), the process becomes technologically challenging. All such aspects of SECC project are explained in chapter 9 by Mahindra. This chapter also demonstrated how multiple demands from various stakeholders, mainly the State governments, were addressed during the SECC processes by modifying and updating the application software housed in these TPCs.

Gender sensitive survey methods are becoming popular these days. When the objective is more than statistical data collection and precisely for identification of beneficiaries for targeting scheme interventions, giving adequate attention to unrepresented sections becomes paramount. The interventions result in redistribution of resources, assets, income and capabilities; and thus, the power balance in not only confined in the society but also within the family. Accordingly, addressing the gender concerns in SECC was made central to the whole exercise, as far as possible. In chapter 7, Kulkarni explains the procedure adopted at each stage of the census to make the SECC project gender sensitive.

SECC is also popularly known as 'Caste Census'. In 2nd week of May 2010, there was an intense debate in the Parliament of India to conduct caste census along with the decennial population census that was due in 2011. Caste population census took place six times in distance past with an interval of 10 years, from 1871-72 to 1931. Since then, no caste census has been conducted. As caste census would quantify the proportion of various castes, the affirmative policy actions as enshrined in the constitution of India can be calibrated properly. That's why it was decided to conduct the caste census along with the decennial population census. However as preparations to add caste census took time, it was decided to collect the caste data at the time of BPL Census (subsequently named as SECC). Instructions to the enumerators were simple—respondent's replay is to be faithfully recorded (realising very well that mistakes may arise due to varying levels of knowledge of the enumerators, data entry operators and the medium in which it is recorded, i.e. TPCs. It was decided to carryout corrections and harmonisation subsequently using appropriate software solutions (similar sounding words). Caste/tribe recording based on the respondents replay only are noted in the SECC 2011 acknowledgement slip issued by the enumerator to the respondents clearly mentioned that "the Caste/Tribe is as declared by the respondent and it will not confer any right to claim any Caste/Tribe/Community certificate." A cryptic review of the history of caste census and current procedures followed to collect caste/tribe data has been explained in chapter 3 by Banerjee and Abraham.

Part 1

Census Operation: Background, Methods, Processes and Practices

1

JOSEPH ABRAHAM

Socio-economic and Caste Census

Beneficiary Identification through Multidimensional Deprivations and Vulnerabilities in India—Capacity Building and Knowledge Management

Introduction

Here in it is proposed to illustrate the voluminous and comprehensive nature of development of knowledge based tasks involved in training, mobilising and managing of personnel involved for conduct of Socio-economic Caste Census (SECC). The training and human resource management functions of SECC have been one of the major tasks addressed by the SECC Team in the conduct of this census. The organisations, agencies and tasks involved in this country-wide exercise are multifarious and have no parallels in the past. Developing of concepts and definitions, issuance of necessary instructions, preparing of training modules, pilot-testing, listing of materials, infrastructures and setting timelines, conduct of trainings that too through multiple agencies, from top to bottom levels of the key stakeholders was a challenging task. Top of the art training technologies, know-hows and skills of talented professionals from among the stakeholders were ensured in the human resource capacity building and knowledge management of SECC. Getting the basic task of obtaining reliable information through canvassing a simple questionnaire in a respondent based e-governance mode via the TPC (tablet personal computer) needed the creation of a friendly ambience and confidence building at all levels of execution of tasks in SECC. Every effort was made to train and orient the formal functionaries and create awareness among respondents in each of the specific tasks mandated upon. A comprehensive narration of the training, human resource mobilisation and knowledge acquisition cum management of SECC as it evolved is attempted here. What is perhaps more illustrious is that the whole of SECC has been done in a digital mode of governance of a paperless census and soft data obtained via use of TPCs for the first time in the country, may be in the world as well.

This Census has covered a total of about 24.4 crore households in the country, of which 17.9 crore were from the rural areas. The basic demographic data from the latest decadal Census 2011 was used as the basic building block on to which socioeconomic data was added and dovetailed into. A total of 2.5 crore of EB (enumeration blocks) were covered in this Census. Per 4 EB one DEO (data entry operator) and one Enumerator was put on filed work to collect information and per 6 Enumerator, one Supervisor was also assigned. The Supervisor was given a separate set of tasks and support of DEOs to carry out the only mandated corrections and those omissions left out by the Enumerators and DEOs were recouped. A total of 6.18 lakh of DEOs plus 6.18 lakh of Enumerators each and 1.03 lakh of Supervisors were mobilised across the country for conduct of the survey. The DVM (data view module) and V&C (verification & correction) functions were carried out only by the government functionaries. The DEOs at EB level and the more techno suave DBAs (data base administrators) at DCCs (data charge centres)—located at each community development block or revenue *tehsil* levels were recruited, trained and deployed across the country by the three CPSUs (central public sector undertakings) namely, Bharat Electronics Limited (BEL) Bangalore, Electronics Corporation of India (ECIL) Hyderabad, and Indian Telephones Industry Limited (ITIL) Palghat. BEL provided technical and manpower inputs for 13 States/UTs (union territories) and deployed 36 venders to execute the technical and statistical tasks, whereas the ECIL took care of 10 States/UTs and requisitioned 31 venders, while ITIL looked after 12 States/UTs and provided the services of 26 venders. These venders were mostly IT companies from the private sector of the country. The government functionaries of the respective nodal departments (rural and urban) performed the supervisory, V&C, monitoring and review of all enumeration responsibilities. In the 640 districts of country, about 10,746 DCCs were set up. For each DCC, one senior district level officer was put in charge by the State/UT governments. In the States, the District SECC Officer was in overall charge of the operational aspects. About 1.60 lakh MTs (master trainers) were mobilised and put to task from among the officials and non-officials to train and handhold the human resources and skill management needs. The cost of the combined SECC 2011 project, which was funded by Government of India (GoI) is pegged at about ₹5,000 crore.

Given this background, the specific objectives of this paper are to: (i) to highlight the differentiated layers of involvement, coordination and collective nature of tasks managed by multiple stakeholders in SECC; (ii) to recapitulate and narrate the resolve to use existing data base from decadal Population Census 2011 as the basic building block and the readiness to use the census concepts with appropriate modifications; (iii) to review and examine the experiment of involving CPSUs as technology and service providers and who in turn recouped the private sector enterprises from the information technology domain for operationalising the e-modules of data collection via data entry operators/data base administrators and its overall management, than fully relying on the governments departmental structures and functionaries; (iv) to highlight the innovative approaches adopted in fixing the various facets of multidimensional poverty; and (v) the training and capacity building approaches and knowledge dissemination involved in SECC are also narrated here.

Convergence among Multiple Stakeholders

The Planning Commission (at present the NITI Aayog) is the nodal agency in the GoI for measurement of poverty. The estimates of poverty were made at the national and State/UT level separately for the rural and urban areas following the methodology detailed in the Report of Expert Group on Estimation of Proportion and Number of Poor. These estimates were worked out by the Planning Commission from the large sample survey data on consumer expenditure carried out by the National Sample Survey Organisation (NSSO) on a quinquennial basis. The Planning Commission used a poverty line based on per capita consumption expenditure as the criteria to determine the percentage and number of persons living BPL (below the poverty line). The Ministry of Rural Development (MoRD) has been implementing various poverty alleviation programmes[1] such as: Swarnjayanti Gram Swarozgar Yojana (SGSY) (now transformed in to National Rural Livelihoods Mission (NRLM) and currently called as Ajeevika), Mahatma Gandhi National Rural Employment Guarantee Scheme (MGNREGS), Rural Housing Programme of Indira Awaas Yojana (IAY) (presently Pradhan Mantri Awaas Yojana (Grameen), Swachha Bharath Abhiyan the former Total Sanitation Campaign (TSC)) and National Social Assistance Programme (NSAP) and its sub-schemes like the Indira Gandhi National Old Age Pension Scheme (IGNOAPS), Indira Gandhi National Widow Pension Scheme (IGNWPS), Indira Gandhi National Disability Pension Scheme (IGNDPS), National Family Benefit Scheme (NFBS) and the recent Sansad Adharsa Grameen Yojana (SAGY). For effective implementation of these poverty alleviation programmes, identification of the rural poor families was called for. Hence, the MoRD had been conducting since 1992 the BPL Censuses through the States/UTs generally in the beginning of the Five Year Plans to identify the rural families who could be assisted under different schemes.

Planning Commission vis-à-vis MoRD on Poverty

In the past, to begin with, the Planning Commission had constituted a 'Task Force on Projection of Minimum Needs and Effective Consumption Demand' which on the basis of a systematic study of nutritional requirements recommended a national norm of 2,400 kilo calories (kcal)/day and 2,100 kcal/day for rural and urban areas (the difference being attributed to the lower rates of physical activity in the urban areas) respectively. These figures were derived from age-sex-occupation-specific nutritional norms by using the all-India demographic data from the 1971 Census. The national-level official poverty lines for the base year (1973-74) were expressed as

1. Name and guidelines of many of these programmes and schemes have undergone changes from time to time.

monthly per capita consumption expenditure of ₹49 in rural areas and ₹57 in urban areas, which corresponded to a basket of goods and services that satisfy the calorie norms of per capita daily requirement of 2,400 kcal in rural areas and 2,100 kcal in urban areas, which was considered as the minimum required for healthy living. The cut-off line has been updated for price rise for the subsequent years.

Hitherto, in India the Planning Commission has been providing the official figures of the percentage of people living below poverty line. And the MoRD has been identifying the poor persons and households through periodic BPL Census. Based on the 61st Round of NSSO results for the year 2004-05 and as per recommendations of the Tendulkar Committee the latest incidence of poverty have been worked out. However, using the Tendulkar and Rangarajan methodologies, the latest estimates of poverty has been estimated by the Planning Commission. These poverty ratios are given in Table 1.1.

Table 1.1

Estimates of Population Below Poverty Line (in %) in India

Year and Particulars	*Urban*	*Rural*	*Total*
1993-94	31.8	50.1	45.3
2004-05	25.7	41.8	37.2
2009-10*	20.9	33.8	29.8
S. Tendulkar Estimates for 2011-12	25.7	13.7	21.9
C. Rangarajan Estimates for 2011-12	30.9	26.4	29.8

Source: Tendulkar Committee Report (2009); *: Revised Estimates, GoI, Planning Commission 2012 and Rangarajan Report on Poverty (Report of the Expert Group to Review the Methodology for Measurement of Poverty, 2014, Planning Commission, New Delhi).

Historical Background of BPL Census

The identification of poor constitutes a critical component of poverty alleviation. The poor families are targeted for central and state assistance in the income and employment generating anti-poverty programmes and also the social welfare programmes. At present, the State governments identify the poor households as per the guidelines issued by the MoRD of the GoI. It is vital to analyse how these surveys were done in the past, and why there was a need to change the methodology every five years? The Integrated Rural Development Programme (IRDP) was launched in October, 1980 in all the development blocks of the country as an instrument of poverty alleviation in the rural areas. Under IRDP, the approach was to assist the poorest among the poor rural families first and the target group was the rural families having annual income below the cut off line of ₹4,800. But families with an annual income up to ₹3,500 were supposed to be assisted first. Registers were maintained at the village level for recording persons identified to be under the poverty line as defined by the Planning Commission in monetary terms; however, there was no systematic survey or census to prepare this BPL list.

The first BPL Census was conducted in 1992 for the 8th Five Year Plan (1992-1997). The need for a systematic survey of BPL families was a felt need based on the feedback revised from the concurrent evaluation of rural development programme that the benefits were given to non-eligible persons and in haphazard ways. During the eight Five Year Plan, detailed procedures were laid down for the identification of BPL families in rural areas. This BPL survey 1992 was conducted using income as the criteria for identification of rural BPL families. The annual family income based poverty line was estimated as ₹11,000 in 1991-92 to separate the poor families from the non-poor. The BPL families were classified into income ranges of ₹0-4,000, ₹4,000-6,000, ₹6000-8500 and ₹8,500-11,000. The identified poor families through the lens of this survey far exceeded the poverty ratio estimated by the Planning Commission.

In the BPL Census 1997 conducted for the 9th Five Year Plan (1997-2002), the criteria for determining the cutoff point was changed from income to consumption. During finalisation of the guidelines for the BPL Survey 1997 for the 9th Five Year Plan, the formal renaming to the nomenclature of BPL Census 1997 was adopted. Before administering the questionnaire to capture the consumption level, an exclusion criteria was applied. The exclusion criteria included threshold level of income of ₹20,000 and above per household, ownership of land, building, consumer durables and agricultural inputs etc. for declaring those families as non-poor. In the second stage, the total consumption, both purchased from the market and home grown, are gathered from the remaining families (who were not declared as non-poor by the exclusion criterion) by interview method. This total consumption of the family is divided by the total number of persons in the family, treating all the members, adult, adolescent and children as identical units. This yielded the per capita consumption expenditure of the family. This per capita consumption of the family is compared with the poverty line (worked out at the state level by the Planning Commission using the Expert Group methodology and used to derive the official estimates of the poverty ratio), and the family is counted in the BPL group if its per capita consumption is within the norm set by the Planning Commission.

The BPL Census 2002 conducted for the 10th Five Year Plan (2002-2007) was based on the methodology recommended by an Expert Group which was constituted by the MoRD. The Expert Group headed by Dr P.L. Sanjeeva Reddy former Secretary of MoRD looked into the issues relating to identification of BPL families in the rural areas and after wide range of discussions, it recommended the methodology of Score Based Ranking of each household indicating the quality of life. For this purpose, 13 socioeconomic parameters including size of landholdings, type of house, availability of clothes, food security, sanitation, literacy, means of livelihood and indebtedness reflecting the levels of living and quality of life of the rural population were identified. The purpose was to make the process of identification of BPL households in the rural areas objective and transparent with the incorporation of mechanism of getting it approved by the Gram Sabha and redressal of grievances. However, in practice the BPL Census 2002 was also criticised as beset with several errors of subjectivity and exclusion/inclusion in identification.

There was considerable delay in finalising the results of BPL Census 2002 as the Supreme Court had passed a stay order on the basis of writ petition No.196 of 20-01 in the matter of PUCL versus Union of India on 5 May 2003. The stay was vacated on 14 February 2006. Till now 27 states and UTs have reported of finalising the BPL list on the basis of BPL Census 2002. Some of the states and UTs such as Kerala, Odisha, Sikkim, Tripura, Lakshadweep and Pondicherry have not yet finalised the BPL List 2002.

No BPL Census was conducted during the 11th Five Year Plan (2007-2012). So it had become imperative to have the next BPL Census done without any further delay.

Processes of Conducting SECC 2011

The Union Cabinet approved BPL Census along with Caste Census in May 2011. Overall responsibilities of the BPL Census in rural areas are with the MoRD, BPL Census in urban areas are with Ministry of Housing and Urban Poverty Alleviation (MoHUPA), and Caste Census are with the Ministry of Home Affairs (MoHA) and MoRD was made the Nodal Ministry for the whole exercise. Registrar General of India (RGI) and Census Commissioner of India worked as an agency of MoHA as the government assured conducting the 'caste census' in the Parliament. The main objectives of SECC are: (i) to make available authentic information on the socioeconomic conditions, and education status of various castes and sections of the population, (ii) to enable households to be ranked based on their socioeconomic status and the State governments can then prepare a list of families living BPL, and (iii) to make available authentic information that will enable caste wise population enumeration.

The MoRD had constituted an Expert Group to advise on the methodology for conducting BPL Census on 12 August 2008. The Expert Group under the chairmanship of Dr N.C. Saxena submitted its report on 21 August 2009. The recommendations of the Expert Group were discussed with State governments/UT administrations, the concerned Central Ministries and other experts.

An expert group was also constituted by erstwhile Planning Commission under the chairmanship of Dr S.R. Hassim to advise on the methodology to conduct the urban BPL Census. Based on the report of Dr S.R. Hashim, the MoHUPA has decided on the methodology of the urban BPL Census 2011.

A Consortium of CPSUs: The CPSUs led by BEL Bangalore had undertaken census work on behalf of Office of the Registrar General & Census Commissioner, India (ORGI) for preparation of National

Population Register (NPR) in coastal areas of the country. It was therefore decided to use the services of this consortium in the BPL Census 2011 also. The other members of the consortium were ITIL and ECIL. The States/UTs were earmarked to them to provide the mandated techno and manpower inputs as per terms of agreements signed with them by the MoRD. The CPSUs have engaged the DEOs and DBAs through their service providing venders.

Role of State/UT Governments

The GoI in June 2011 had issued detailed instructions for the conduct of SECC 2011. However, it is to be noted that as the proposed census goes beyond BPL Census, by collecting 'caste' data also, it was named SECC. As per these instructions, the SECC 2011 will be carried out by the respective State/UT governments with the technical and financial support of the GoI and the notifications to be issued by the appropriate authorities in the gazette. Besides making the materials available to States/UTs by postal mailing, the same 80 page document is made available in the MoRD website (*www.rural.nic.in*). These were to be used as a guide by the various stakeholders for the conduct of SECC 2011 with necessary modifications as per State/UT requirements. Some of the important responsibilities of the States/UTs were the setting up and running of DCC, appointment and training of Enumerators/Supervisors/Charge Officers/Census Staff, conducting of enumeration, data view and supervisory verification and error corrections, draft publications and claims & objections.

Role of ORGI: The structure of SECC is solidly built on the strong pillars built up by the ORGI and the Census Commissioner of India in the conduct of the decadal Censuses in the country. An array of the strong building blocks of SECC is provided by the ORGI. A few of these being the NPR and house-listing data base, enumeration blocks, abridged house lists, house numbering, layout maps, the concepts and definitions used and in streamlining the conduct of the Caste Census related issues and to prepare socioeconomic profile of various castes etc. The unstinted support of the DCOs (directors of census operations) of each of the State/UTs has been another noteworthy input from the ORGI for the smooth conduct of SECC. Besides, a team of dedicated officials of the Census department had played crucial roles in each stage of the conduct of SECC.

National Informatics Centre (NIC): The NIC had been working together with the MoRD since several years and being engaged in developing the MIS (management of information system) for many programmes. Thus NIC was engaged to develop a web based MIS for the management of the database of SECC 2011 and to facilitate its subsequent use by the programme divisions, other ministries and State governments for their own requirements. The restricted MIS site is: *https://sec-census.nic.in*. The MoRD website *https://rural.nic.in* also provides details of the SECC 2011 guidelines, orders and formats etc. The results of SECC were displayed in public domain developed by the NIC. The draft and final lists shall be displayed on the website *https://secc.gov.in*, for access by public. Lastly, the dynamic and analytic reports based on final SECC database[2] could be generated by authorised persons from *https://secc2011.nic.in*.

Regular review and monitoring: The progress of conduct of SECC was reviewed and monitored through periodic video conferencing and visit to States/UTs. State/UT specific reviews and assessment of progress of work was also undertaken. Based on review and decisions in the Performance Review Committee—a committee comprising the senior officials of MoRD and State government secretaries' in-charge of SECC project—the senior officials from MoRD visited some of the States to expedite the census process. For the northeastern states, region specific meetings were also held to review progress. The Chief Managing Directors and General Managers of the CPSUs and the heads of stakeholder departments (ORGI/MoHUPA/NIC) were also been involved in the review meetings conducted by the Minister, Secretary, Chief Economic Adviser, Additional Secretary and senior officials of MoRD. Senior officials of MoHUPA and Office of RGI and Census Commission of India also monitor the exercise and provide necessary directions as and when required.

2. SECC database management and operation of MIS is covered in Chapter 11.

Data Quality Check

During each of the major steps of data collection efforts, built in quality checks were envisaged. The service providers were required to ensure the quality of data. Wherever gross errors were observed, remuneration was to be conducted. The numerous mismatches observed; in the numbers and codes of EBs, in the schedules used and duplication of data identified in the process have been communicated to the States/UTs and CPSUs to rectify the errors.

The utmost cooperation of DCOs of the States/UTs have been received in settling a number of issues with regard to mismatch/errors in the NPR-TIFF Images, in the number of EBs and sub EBs, and in the enumeration of special chagres.

The Rural and Urban Nodal departments of the State governments and UT administration have always been working to achieve data quality not only by ensuring quality output by the Enumerators and Supervisors, but also by insisting that the quality of data must be ensured by the active support of the CPSUs and the venders. Managing and handholding the venders/service providers to ensure quality data have been a major challenge for the CPSUs and nodal departments.

Time Over-run, delays and cost escalation: The convergent and multidimensional approaches, different timelines planned and taken up, various stage-wise procedures and new processes that got rolled out in the conduct of SECC are unique as these are attempted in democratic, bureaucratic, technocratic ways, together with civil society cum media support, and in participative manner and all of these in an e-governance mode. Though in the conduct of SECC, the time lines got extended and costs escalated, many novel experiences were gained in the entire gamut of conduct of activities. These valuable experiences could be remoulded and put to more effective use in future rounds of BPL and other beneficiary Censuses in India and other countries.

Built on strong foundations for digital-census: The ORGI and the Census Commissioner of India has provided for logistic and technical support including the following:

a) To ensure complete coverage and smooth conduct of SECC 2011, the directory comprising of complete location codes and names of State, district, sub-district, villages/town, and enumeration block/sub-block used in population enumeration phase of Census 2011, photocopies of abridged house list, charge register and layout maps have been provided;
b) Scanned images of the household schedules of NPR canvassed in the year 2010 for updating through the handheld devices used in the SECC;
c) ORGI monitored the progress of SECC in States and UTs with regard to coverage of population and caste/tribes being reported from the field;
d) ORGI imparted training to various functionaries on census concepts as also on specific issues concerning 'religion' and 'caste/tribe';
e) Quality checking of SECC data with particular reference to religion and caste/tribe has been taken up by ORGI;
f) Test data of selected fields from the NPR for replacing NPR image data in the SECC database; and
g) After data collection from the field, the ORGI would process the caste/tribe data and handover the same to the proposed Expert Group to be constituted by GoI by including the Ministry of Social Justice & Empowerment and the Ministry of Tribal Affairs to classify these caste/tribe returns.

Cataloguing Procedure

These refer to processes for transferring scanned images of name of head of households obtained during the first phase of Census 2011 canvassed i.e. during the House-listing Operations held from April to September 2010 were obtained from ORGI and uploaded into handheld device/TPC. The application software developed helps in executing this task and in establishing the link to the household of the respondents at the time of survey. In case, the images are not present, a dummy will be created by the software. The procedures for house numbering for new households that came into existence after

Census enumeration were also illustrated and familiarised in the training sessions (the details are given in the Enumerators Manual).

CPSUs, Public-Private Sector IT Service Providers and Civil Society

Role of NIELIT: National Institute of Electronics & Information Technology (NIELIT), (erstwhile DOEACC Society), an Autonomous Scientific Society under the administrative control of Department of Electronics & Information Technology, Ministry of Communications and Information Technology, GoI, was assigned the task of digitising NPR data for merging with SECC data.

Project Management Unit (PMU) of the National Institute of Smart Governance (NISG): The PMU was set up to provide strategic inputs and administrative support to MoRD in the planning and implementation of BPL Census 2011. It provided general status update to the MoRD and other stakeholders on a regular basis. The organisation of meetings, video conferences and providing of SECC information updates is also being done. The PMU also interacts and coordinates with MoRD officials, consultants, other Ministries and CPSUs to generate updates of the current developments and provide valid feedback.

The UN Women and SECC 2011: United Nations (UN) Women has given manpower and technical input in the preparation of IEC materials for SECC besides placing full time services of a gender expert. The SPARSH/TARAA NGO consortium organised one training for a select group of six NGO representatives across the country and the same was organised under the auspices of United Nations Development Program (UNDP). The participants of this training were from across the country and were requested to focus on imparting training to the filed level functionaries by coordinating with the respective State/UT governments earmarked to them in the conduct of successful dissemination of key knowledge and awareness building for SECC 2011.

Independent Monitoring Agencies (IMAs):[3] Though the system put in place to conduct the SECC 2011 aims at ensuring transparency,

3. In chapter 6 of this book, the challenges in conducting such an exercise is explained.

accountability, public-participation, inclusiveness and authenticity of the data collected, the government wanted to be doubly sure to arrive at the desired outcomes of this census operation. A Committee with Dr N.C. Saxena, Member, National Advisory Council, New Delhi as Chairperson, has been constituted to advise on the procedure to setup a system/mechanism comprising of institutions/persons that would closely supervise and monitor the conducting of the SECC so that the census operation is subjected to rigorous scrutiny by third party independent agencies and civil society.

The objectives of the Concurrent Monitoring System are to engage third party independent agencies to concurrently monitor the processes of SECC 2011 as mentioned in the guidelines and ensure: (i) whether data is being collected accurately and fairly; (ii) household coverage irrespective of their socioeconomic status; (iii) accuracy of the data being collected; and (iv) transparency, inclusiveness, justice and fairness achieved while the data/information is collected/finalised before it is put to use. The target for concurrent monitoring process of SECC is to cover at least one per cent of all EBs in rural areas to ensure that the data has been collected accurately irrespective of the socioeconomic status of the households. (In chapter 8 of this book, the challenge in conducting such an exercise is explained.)

Innovations to Illustrate Multidimensional Poverty

The Expert Group to Advice on the Methodology for BPL Census

The MoRD constituted an Expert Group to advise on the methodology for conducting BPL Census in August 2008. The Expert Group under the chairmanship of Dr N.C. Saxena submitted its report in August 2009. The recommendations of the Expert Group were discussed with State governments/UT administrations, the concerned Central Ministries and other experts.

Pilot study: Based upon the consultations with Planning Commission, experts, representatives from various Ministries and State governments, it was decided to conduct a Pilot survey to field test alternative criteria for exclusion and inclusion in order to arrive at the appropriate methodology for BPL census. It was also decided that the inclusion and exclusion criteria of the Saxena committee would be used as reference points for the pilot along with other criteria for exclusion and inclusion recommended by experts, states etc. The Pilot would also enable testing various combinations of criteria which can be used to identify BPL families. The MoRD set up a Core Group comprising of experts and representatives from the Planning Commission, ORGI, NSSO, Unique Identification Authority of India (UIDAI) and National Advisory Council (NAC), a Central University and others to oversee and guide the process of the Pilot and the BPL Census.

Findings of the pilot study: On the basis of suggestions of the experts, a pre-testing of the methodology through a Pilot Study was conducted. Results of pilot study were discussed with experts and States/UTs in February 2011. The Pilot survey revealed that specific automatic exclusion indicators would possibly exclude approximately 28 per cent of the rural households. The pilot also identified a list of deprivation indicators. It is estimated that use of three or more deprivation indicators (SC/ST treated as one deprivation) would include approximately 15 per cent of the households and use of 2 or more deprivation indicators would include approximately 35 per cent of the households at all-India level.

The findings of the pilot study were shared with the core group of experts in a meeting held in February 2011. The Core Group endorsed the findings of the pilot study and undertook careful consideration of different methodologies and advised the MoRD to take a view on the subject. A meeting with State Secretaries of Rural Development was held in February 2011 in New Delhi to discuss the findings of the pilot study. Representatives from the State governments broadly agreed with the findings and suggested that criteria such as landless agricultural worker, primitive tribal groups, physically handicapped, and chronic illness may be considered for inclusion. They further suggested use of State-wise indicators to capture the peculiarity in the States as also district-wise and block-wise poverty caps and adoption of scoring methodology to generate ranking of the poor.

SECC 2011

The GoI had taken a decision to conduct a Caste Census for the country and assurance to this effect has been made before the Parliament. The BPL Census is also required to be conducted for the urban areas in order to provide basis for the urban poverty alleviation programmes. Conducting these three Censuses separately would have resulted in duplication of efforts, wastage of resources and also contribute to Census Fatigue (CF) over the country at large particularly in wake of the preparation of the National Population Registers and the Census 2011 in quick succession. In order to prevent the afore mentioned duplication of efforts and times, the GoI decided that the three Censuses as mentioned above would be canvassed together in a convergent mode with the MoRD acting as the Nodal Ministry. Hence the SECC 2011 was launched in November 2011, with Tripura being the first State selected for the purpose.

Methodology of the BPL Census 2011 (Rural)

Based on consultations with the States/UTs and other experts, in the light of the findings of the BPL pilot survey, the proposed methodology to conduct the door to door census of rural households for identification of BPL households is as below:

a) The Census will be conducted by State governments/UTs with the technical and financial support of the GoI. The Census would be based on a self-declaration model of the respondents. The Enumerator would record the declaration of the respondents and where he finds that the response is prima facie incorrect, he would record his observation separately in the space provided for in the enumeration format. The Supervisor would mandatorily visit these houses and verify the veracity of these set of information. The Supervisor would also monitor and randomly check the filled up questionnaires on a sample basis based on a set norms. The collected information would also be verified and approved by the Gram Sabha. Procedures for filing claims and objections and its disposal will be prescribed separately by the Ministry. The Ministry will also provide detailed guidelines for conducting survey and finalisation of BPL list. The survey method, the schedules designed for rural and urban areas was finalised before the start of SECC. However, the original schema of identifications of BPL/beneficiaries conceptualised that the classification of households would be done in three steps in rural areas. It is explained in subsequent paragraphs.

b) *Exclusion criterion*: The exclusion criteria would be that any household satisfying any of the following criteria would be excluded:

   i) Households owning motorised two/three/four wheelers/fishing boats (which require registration);
   ii) Households owning mechanised three/four wheeler agricultural equipments such as tractors, harvesters etc;
   iii) Households having Kissan Credit Card with the credit limit of ₹50,000 and above;
   iv) *Households with any member as Government employee*: gazetted and non-gazetted employees of Central government, State government, public sector undertakings, Government-aided autonomous bodies and local bodies. This will exclude incentive and other honorarium based Asha, Anganwadi workers etc.;
   v) *Households with enterprises registered with the Government for any purpose*: any non-agricultural enterprise registered with the Central or State governments;
   vi) Households with any member in the family earning more than ₹10,000 per month;
   vii) Households paying income tax or professional tax;
   viii) Households with three or more rooms with *pucca* walls and *pucca* roof;
   ix) Households owning refrigerator;
   x) Households owning landline phones;
   xi) Households owning 2.5 acres or more irrigated land with at least one irrigation equipment such as diesel/electric operated bore well/tube well;

xii) Households with 5 acres or more land irrigated for two or more crop seasons;

xiii) Households owning 7.5 acres or more land with at least one irrigation equipment such as diesel/electric operated bore well/tube well;

c) *Compulsory inclusion*: Following categories of households would be compulsorily included:

i) Households without shelter;

ii) Destitute/living on alms;

iii) Manual scavengers;

iv) Primitive tribal groups; and

v) Legally released bonded labourers.

d) *Deprivation criteria*: The remaining households (subject to exclusion criteria and compulsory inclusion) will then be assigned deprivation scores depending on the number of deprivation indicators they satisfy. The following are the deprivation indicators used for inclusion:

i) Households with only one room with *kachha* walls and *kachha* roof;

ii) Households with no adult member between age 16 to 59;

iii) Female headed households with no adult male member between ages 16 to 59;

iv) Households with any disabled member and no able bodied adult member;

v) SC/ST households;

vi) Households with no literate adult above 25 years; and

vii) Landless households deriving major part of their income from manual casual labour.

e) *Inclusion priority*: Households eligible under compulsory inclusion will have highest priority. For the households eligible for ranking under deprivation indicators as per above paragraph, a deprivation score would be derived for each household by adding up the number of deprivations satisfied by the household. This score will vary from a minimum 0 to maximum 7. The order of priority for inclusion of households in the BPL list would be from largest number of deprivations to smallest number of deprivations. For the purpose of coverage under welfare schemes of the Government, households eligible for compulsory inclusion will have highest priority, followed by households with higher deprivation scores.

Joint Statement by the Minister for Rural Development and Deputy Chairman Planning Commission: On 3 October 2011 in a joint statement made by the Hon'ble Minister for Rural Development and Deputy Chairman Planning Commission, the following clarifications have been made on the outcomes expected out of SECC. Specific entitlements that rural households will receive under various central government programmes and schemes be decided after the SECC survey results are available and analysed. The present state-wise poverty estimates using the Planning Commission methodology will not be used to impose any ceilings. Consultation with states, experts and civil society organisations will be done to arrive at a consensus on the methodology by the time the SECC 2011 is completed. An Expert Committee will be appointed to ensure that this methodology is consistent with the provisions of the Food Security Bill as it finally emerges.

All in all E-Modules Mode of Governance: The entire gamut of activities of SECC was envisaged and executed in a digital mode. For each stage of specific activity, separate software is developed and put in use. The client and server versions of software are used for enumeration, supervision. The verification and correction, data view module and COTS (claims and objections tracking systems) and the Final List Generation are also placed at the data charge centre/ district levels. The meaning and usefulness of these modules are discussed below in detail.

Use of Handheld Enumeration Device for BPL Census 2011: Use of low cost handheld device for collection of data in BPL Census 2011 was also discussed with the representatives of the State governments, who supported its use as it would reduce the time required in processing the data after collection in the field. Thereby it was decided to use this device, produced by BEL, in SECC. The portable hand held device or TPC developed by BEL is extensively used in the SECC. It was envisaged as a paperless e-survey.[4]

Various Processes of SECC

Canvassing of the BPL and Caste Census: It was decided that the BPL Census would be conducted along with Caste Census being conducted by the office of ORGI and urban BPL Census to be conducted by Ministry of Housing and Urban Poverty Alleviation (MoHUPA). As required by ORGI for Caste Census, information related to caste would be separated after the data collection was over and handed over to ORGI for further processing. Information on the caste names would not be draft published.

Concepts and definitions: The concepts, definitions and designs used for SECC are defined and are provided in the manuals circulated for the enumerators, supervisors and V&C officers, and the COTS operators.

Questionnaires used: An electronic questionnaire was used for data collection for rural and urban areas. Side A and B of the questionnaire collected individual and household specific information respectively. Side B of rural questionnaire differs from that of Side B of urban questionnaire. In the questionnaire, Columns 2 to 9 (Block B) was common format and contained information from the NPR houselisting data for both the rural and urban areas separately. Such houselisting data served as base and subjected to corrections and then allowed updating. Block C includes religion and caste/tribe status.

Pre-enumeration activities: The identification particulars and codes of the EBs, the layout map and the abridged house lists of each EB and how to set up the DCC and what facilities are to be provided are all informed to the State/UT governments. The Enumerator's Manual containing all the detailed instructions was issued to the State authorities and made available in the website.

Major steps in SECC 2011: Each of the major steps of SECC being data collection/enumeration, supervision, data view, verification &

4. Manufacturing and use of TPCs is covered in chapter 8.

correction, draft list publication, settling of claims and objections in the Grama Sabha (and ward sabhas in urban areas) and release of final list in the NIC website; were all conducted through the application of relevant software developed by the BEL and NIC. These versions of the softwares were developed through a process of trail and updating.[5]

Software being used in SECC 2011: The SECC data collection is in total a digital effort. And each stage of SECC work is implemented with the support of different application software. These software are developed by the BEL and NIC. Before putting to actual use, trial runs were undertaken of each of these software and feedback received and the field encountered practical difficulties were taken into consideration and appropriate modifications incorporated. Therefore different versions of the software were put to use in the field. Some of the software versions used and the particulars of each are the following: data entry, enumeration and supervision (client and server); draft list versions; DVMs; V&C modules; COTS versions; and final list generation.

MoRD has circulated guidelines which include pre-enumeration, enumeration and post-enumeration activities in detail (for details see the web sites: *http://rural.nic.in/bplcensus_page.htm* and *http://rural.nic.in/circular/Guidlines_27062011.pdf*). Similarly, MoHUPA issued circulars and guidelines on SECC for urban areas (*http://mhupa.gov.in/User_Panel/UserView.aspx?TypeID=1526* and *http://mhupa.gov.in/User_Panel/DetailsView.aspx?TypeID=1219&ID=1406*).

Enumeration: The Enumerator had to visit each household along with the DEO and get the details entered into the TPC. The questions will be asked by the Enumerator and the responses will be entered into the TPC by the DEO.

Supervision: The Supervisor has to perform well defined supervisory functions through the TPC with the help of application software and DEOs. A minimum of 10 per cent of the enumerated households per EB is mandated for supervision. However, the types of households selected for mandatory supervisory verification

5. Details are in chapter 9.

are the following: (i) all unenumerated households along with "reason for non-enumeration"; (ii) all unenumerated catalogued households' image files as per AHL Section 2, 3 and 4 of catalogued households; Enumerator disagrees with the information furnished by the respondent of a household; (iii) caste/religion information missing in the relevant boxes; the household's whose data collected by enumerators but as selected via the use of 'DVM'; and (iv) lastly if all the above types of households do not add up to 10 per cent of the total then the deficit only will be made up from among the good records randomly.

Data View Module: To facilitate the supervisory functions, the DVM was introduced so that in the DCC the authorised officer can select the erroneous household returns for supervision. The DVM enables the DCC in charge to view all the data collected by the enumerators. This facility enables the Charge Officer to assess the quality of data collected. Wherever the correctness and authenticity of data collected was doubtful, such households (e.g. large number of legally released bonded labours than the official figures with the administration, religion and caste information misleading etc.) can be brought under the purview of the scrutiny by the Supervisor. Those households flagged by the Charge Officer for supervisory field check will become part of the e-files for supervision. Only such files were to be edited by the Supervisor and remarks entered in the relevant column. The Supervisor was to be assisted to a DEO while the house to house visits are done.

Verification and Correction Module (V&C): The V&C module was introduced to ensure the quality of data with respect to a specific set of parameters. The Verification Officers will have to physically verify the mandated fields namely the PTG (primitive tribal groups)/LRBL (legally released bonded labourer)/MS (manual scavenger) in the rural areas and for the urban areas the four types of major sources of income from rag picking, begging, domestic jobs and street venders. A select list of common errors in reporting the caste names/codes have been identified by ORGI and put to verification and correction. The files for verification will be in PDF format and with relevant space obtaining the signature of the respondent and for writing the remarks of the verifying officer. No TPC is used at this stage as verification is done on paper schedules. An additional caste correction module also was introduced to take care of errors noted even after execution of V&C module.

The need of V&C module has become obvious for the following reasons. Form all most all the States and UTs, the enumerated data have thrown up the following challenges. Firstly, a disproportionally very high number of households reported to have members belonging to PTG, LRBL and MS. Secondly, in urban areas, data collected on the numbers of occupational categories like beggars and rag-pickers are found to be unreasonably high. Thirdly, an apparently wrong/ incomplete returns of caste/tribe name, e.g., 'other'/'open'/ 'general'/'OBC'/abbreviated caste name/ religion name, etc. have been found in enumerated data.

In the implementation of V&C module all the States/UTs are advised to implement this module before draft list publication. With respect to the above mandated fields, the household numbers with member details, EB numbers with their reported/enumerated data would be provided at the DCCs with the help of application software designed by the NIC and BEL. These will be printed form for each of identified households with earlier entry and space for new/verified entry. The State government will send one official designated as the 'Verifying Officer' to those particular households and get the correct data from respondents.

The corrected data will be entered and uploaded at the DCC with the help of the application software and will be saved in the NIC server. The CPSUs will circulate the instruction manuals for the application software and organise training for the DEAs.The service provider CPSU would ensure the engagement of DEAs in required quantity and quality.

Draft list publication: SECC draft lists to be published in the local language contains basic detailed information on the individuals and households and it will be displayed in the public domain for undergoing the claims and objections procedures.

Digitisation of NPR data and matching with SECC: The first step in the 'Claims & Objection' procedures is to display the Draft List containing the household data collected. The demographic features of the household (member name, relation with the head, gender, and date/year of birth, fathers/mothers name, marital status, and occupation/activities) in the draft list are in image format and the rest of data is in text format. This is because during the SECC, selected portions of the NPR images were used as a base in the TPC while collecting the field data. At a later stage it has been decided to present all the draft list data in text format. It was evident that the cropped images were not legible or readable by the general public besides it was not amenable to analysis and hence it was decided to digitise such images data. Taking into consideration the long time required for converting the unmatched household images, it was decided to publish the Draft List with both the machine converted matched images of household members in text format and the remaining unmatched household images the name of the household members will be cropped up from the images and pasted in the draft list. The CPSUs are put to the task of undertake the machine matching. The process of 'manual matching' can also be done parallel with the Claims and Objection process. The CPSUs have been directed to undertake the machine and manual matching and provide the necessary services for Draft List generation. The Draft List generated after digitisation of NPR-SECC data can have both text and cropped images together. The percentage of cropped images in the draft list after digitisation and merging was shown to be very low at 15 per cent as per estimates.

Protocols for Draft List generation: Herein, the mandatory tasks to be completed are listed for upkeep by States/UTs before generation of CSV (coma separated value) files for merging the NPR and SECC data. The former tasks are completion and uploading of enumeration, supervision, V&C data files and settlement of jurisdictional errors. Some of the other errors noted are duplication of EBs and zero population uploads etc. Once the State/UT governments comply with the detailed protocols of the above four steps, the Nodal Officer will send an intimation to NIC and MoRD that the process for NPR text merging can be initiated. The NIC will check for upkeep of all mandatory protocols for particular districts from States. Then, after getting the NPR digitised files from the local DCO and the SECC*. CSV files from NIC, the CPSUs will merge the same. These merged database files will have to be uploaded by CPSUs to NIC Server at local NIC office. And only for the States/districts where the NPR text merging has been completed by CPSU and data uploaded/re-uploaded, the NIC will generate one time incremental CSV for NPR text merging. Finally, the States/UTs have to give a declaration to MoRD that all preparatory works for generation of draft list have been completed. Then MoRD will request the NIC to generate the data base file for draft list generation. MoRD has issued the protocols to be maintained by the States/UTs in getting the draft and final lists printed.

Protocols for jurisdictional error corrections: This refers to errors occurring in the location parameters used to indicate rural and urban status of the enumeration blocks. A rural enumeration block may have been enumerated as urban and vice versa. Jurisdictional errors may have taken place in part or entirely for an EB. In both cases re-enumeration and/or re- supervision or both will have to be done depending on the nature of error. EBs with zero population upload and those un-inhabited EBs will have to be identified by the State/UT through the local DCO and corrective action taken. Necessary protocols have been issued for identifying and sorting out the jurisdictional errors.

Claims and Objections settlement: The Claims and Objections settlement procedures and timelines have been laid down to enable the respondents to call for changes, deletion and additions of relevant information. The placing of the Draft List in the Grama Sabhas of the Grama Panchayats has been envisaged for rural areas. The appellate mechanisms have also been spelled out. A web based software, called the COTS has been developed by BEL and is put to use. The SECC final list will be uploaded in the NIC enabled website of *secc.gov.in*.

Procedures for claims and objections: In order to promote transparency, verifiability and accountability of the data collected, a system of Claims and Objection settlement procedures have been laid down for SECC 2011.

Public notice prior to draft list display: At the time of data collection, the respondents were to have been briefed of their consent or otherwise on placing the data collected in the public domain except that of the caste name and religion. A signed acknowledgement slip was issued by the Enumerators after canvassing to the respondent. The States/UTs were requested that a public notice be issued in the media well in advance that the SECC data would be put in public domain unless the households submit a written request in the prescribed format that they do not wish to share their data in public. A list of such households without household details will be separately appended in the Draft List of each enumeration block. If they do not submit such a written request within 15 days of the issue of media notice, then it would have been presumed that there is no objection to share the data in the public domain and proceeded accordingly.

The major features of Claims and Objections on the draft list data published are the following:

A. Use of prescribed forms by petitioners and designated officials.
   - i) Form A is used for filling objections against inclusion of particular data;
   - ii) Form B for making corrections and modifications;
   - iii) Form C for inclusion of data;
   - iv) Form D for issue of notice to the objector/respondent;
   - v) Form E for recording the proceedings of the Grama Sabha;
   - vi) Form F for marking the respective corrections done at the time of disposal of claims. A duly filled and signed copy of form F has to be given to the petitioners. The Designated Officer for settling the Claims and Objections will be appointed by the State/UT Governments. The Appellate Officers to settle appeals that come up over the decisions of the Designated Officers will also be appointed by the State/UT authorities. The timelines for the various stages and procedures of the entire claims and objections processes have been clearly earmarked and communicated to the States/UTs.

B. Claims and Objections Tracking System: The COTS software provides for the web based online maintenance of information of the procedures and decisions taken up on each of the claims and objections on the draft list SECC.

C. Final list to be displayed in the website: Once the COTS procedures are successfully completed, the final SECC list will be put on the website as per adherence to the timelines prescribed and kept by the States.

Administrative arrangements: The State/UT Governments of Rural/ Urban/ DCO departments are the nodal functionaries for conduct of SECC. The Principal Secretary for Rural Development Department is the overall in charge at the State/UT level. The basic infrastructure for the conduct of the SECC will have to be provided by the State/UT governments. The enumerator/supervisor/verifiers are appointed and put to service by the States/UTs.

The respective CPSUs assigned to each State/UT has the task of providing the basic technical services, will also have a State Nodal Officer to coordinate the activities with the governments at State/ District/DCC and at block/*tehsil* levels. The DEOs, DBAs at the DCC levels will have to be recruited, trained and positioned by the concerned CPSUs as per the terms of agreement with the private venders in consultation with the States/UTs.

COTS have been developed for urban areas too with minor modifications to address the administrative dissimilarities in the field level.

Constitution of Expert Committee on SECC 2011: On December 2012, in pursuance with the commitment made earlier jointly by the Deputy Chairman Planning Commission and the Minister for Rural Development, the MoRD constituted an expert committee with the following composition: (i) Prof Abhijit Sen, Member, Planning Commission (Chairperson), (ii) Dr Kirit S. Parikh, RIDe, (iii) Prof Mahendra Dev, Director, IGIDR, (iv) Shri Nikhil Dey, Suchana Evum Rozgar Aadhaar Abhiyan, (v) Dr Himanshu, Jawaharlal Nehru University, (vi) Dr C. Chandramouli, RGI and Census Commissioner, GoI, (vii) Dr Indu Agnihotri, Director, Centre for Women's Development Studies, and (viii) Shri P.P. Mitra, Chief

Economic Adviser, MoRD. The committee examined the rural SECC indicators and conducts the data analysis to recommend appropriate methodologies for determining classes of beneficiaries for different rural development programmes. It will consult States, Experts and civil society organisations while arriving at these methodologies. The committee submitted its report to MoRD in May 2014.

Human Resources and Skills

Training Strategy a Cascading Model

It was initially realised that training of all levels of functionaries is a must for successful conduct of SECC.The written instructions on soft and hard forms on every aspect of conduct of census was made available to all stakeholders. The services of experienced trainers across the country were made available across the country. The capacity building and knowledge dissemination management of SECC was undertaken on a structured manner. The major planks of it are narrated below (see Figure 1.1).

Provision of effective training of field functionaries is a pre-requisite for successful conduct of a nation-wide census. These involved preparing written instructions and manuals. The services of experienced trainers were requisitioned. An array of NLTs (national level trainers) at country level and MTFs (master trainer facilitators) and MTs at State/district and DCC levels were identified, trained from among the DCO/Nodal Departments/NGOs/ BNVs (bharat nirman volunteers) and CPSUs and put into action.

NLTs for each State and UT were given training at National Institute of Rural Development (NIRD) and in New Delhi. From each State/UT, a representative number of senior officials from the nodal departments namely; Census, Urban Development, Rural Development, Panchayat Raj and from there CPSUs were given intensive training on the conduct of SECC. It was instructed that these NLTs will in turn train sufficient number of MTFs & MTs at State and District levels and the latter in turn will impart training to DCC officers, supervisors, Enumerators and DEOs and DBAs. The involvement of NGOs/civil society members and BNVs in the pool of resource persons as MTs was encouraged. These trainings were to be done in joint efforts with the CPSUs as the latter will have to train the DBAs and DEOs with the required technical inputs. It was also instructed that the trainings conducted for enumerators and Supervisors would also be for one days/session of joint training with the DEOs/DBAs being requisitioned by the CPSUs and venders. Details of training conducted for NLTs is given in Table 1.2.

Training materials/modules in English and Hindi both in hard and soft copy forms were made available to the States and UTs. The local language versions of these materials were to be made and locally supplied by the State/UT governments. The manuals meant for exclusive use by the enumerators, supervisors and verifiers were prepared by ORGI, MoHUPA, NIC, MoRD and NIRD & PR. Figure 1.1 gives a bird's eye view of HR, CB and knowledge managements involved.

State specific trainings and field visits were also arranged in some of the large states and in problem areas.

The training modules covered the following subjects: (i) Updating of abridged house list/layout maps/numbering of buildings and census houses; (ii) Use of TPC and various software applied; (iii) Concepts and definitions of SECC; (iv) Identification particulars; (v) cataloging of households; (vi) NPR schedules & abridged house list operation; (vii) Questionnaire and the definitions of the parameters used (viii) Mandatory declarations; (ix) Roles and responsibilities of data entry operators, enumerators and supervisors; (x) Data view module; (xi) Verification and correction module; (xii) Draft list publication; (xiii) Claims and objections procedures; (xiv) Additional caste names error corrections; (xv) Final list generation protocols; (xvi) Gender concerns; (xvii) Role of NIC and MIS applications; (xviii) role of CPSUs (and its venders)/state/UT governments; and (xix) use of TPCs; (xx) and role of NGOs/civil society/BNVs. These were made available in soft copy and printed formats.

IEC strategy: The guidelines issued by MoRD clearly specified that the information and awareness generation on the SECC process must be carried out through newspaper advertisements, billboards, radio spots and television coverage. IEC (information communication and education) strategy was developed and material designed by MoRD and disseminated to the various States/UTs for further use in local language versions.

It was proposed to have intensive campaign of about two weeks in the early stage of the Draft List Publication and Claims & Objection stage. The campaign would be in local/state languages.

Posters: A set of six posters focusing on the need to provide the correct information, facilitate participation of women, and reporting various forms of disability etc for SECC was prepared. The soft copy of these set of posters was provided to states and they were requested to use it in their local languages to give wide publicity in the SECC campaigns mandated upon. These were prepared with the support of UN Women of the UNDP, India unit.

Video spots for TV: A 30 seconds video spot was made and circulated. It highlighted how the people can voluntarily participate in SECC and the use of TPCs. These were prepared with the active support of SIRD and NIRD at Hyderabad.

Radio spots: A 50 seconds audio spot was prepared with the help of the National Film Development Corporation (NFDC). The audio spot used the mainstream cinema character voice modulating styles to attract the listeners. It aimed at emphasising the providing of right and timely information to SECC enumerators by the respondent public.

Toll free call centres: A toll free call number 1800-180-2012 has been set up with the technical support of BSNL (Enterprises Business) Faridabad. The personnel of the BSNL were trained for the purpose. 14 regional centres have been set up where local language based assistance has been provided to the people on SECC related queries.

The PRIs (panchayati raj institutions) have been involved in the conduct of SECC. The draft SECC List will have to be prominently displayed in the Gram panchayats. The Gram Sabha will have to be called up by the *panchayats* for scrutiny of the draft list. The State/UTs have been requested to involve the PRIs in the IEC campaign for SECC.

Figure 1.1

Human Resources: Training and Mobilisation in SECC

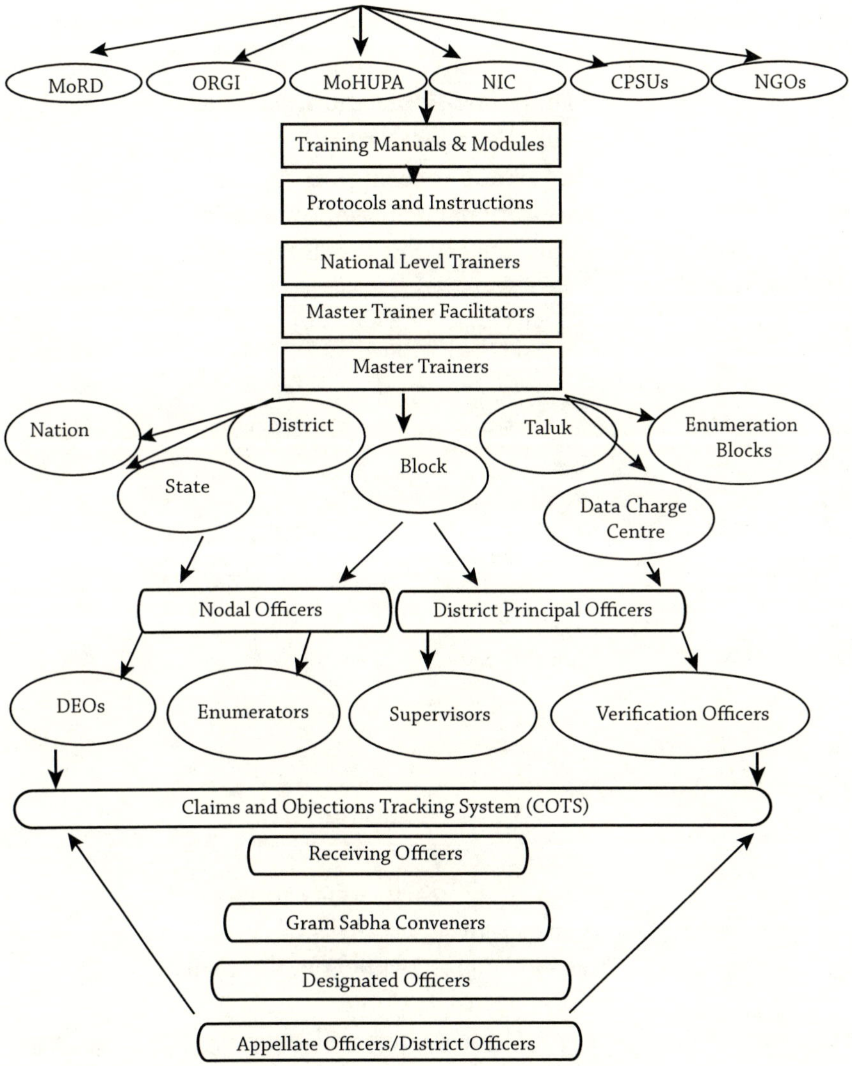

Within 10 days of draft list being published, a Gram Sabha meeting, will be called for public scrutiny in rural areas. In this meeting a representative of district magistrate/collector shall remain present. The Gram Sabha shall scrutinise the Draft List for all the

Table 1.2

National Level Trainings (NLTs) Held on SECC 2011 by MoRD Training Team

No.	*Dates*	*Venue*	*Topic*	*No. of NL Trainees*
1	24-26 July 2011	NIRD Hyderabad	SECC to National Level Trainers (NLTs)	62
2	23-26 Aug 2011	-do-	-do-	71
3	6-9 Sept 2011	-do-	-do-	84
4	20-23 Sep 2011	-do-	-do-	42
5	18-21 Oct 2011	-do-	-do-	61
6	10-11 Feb 2012	At NASC Complex New Delhi	Supervisory Manual and COTS to NLTs	65
7	16-17 Feb 2012	-do-	-do-	62
8	27-28 Feb 2012	NIRD Hyderabad	Supervisory Manual and COTS	44
9	29 Feb-1 Mar 2012	-do-	-do-	49
10	31 Oct-2 Nov 2011	At Vishwa Yuvak Kendra, New Delhi	SECC Orientation to Trainers Alliance NGOs Grid through the UNDP	90
11	5-7 Nov 2011	-do-	-do-	90
12	10-11 Feb 2012	NASC Complex, New Delhi	Supervisory Module and Claims and Objections Processes	75
13	16-17 Feb 2012	-do-	-do-	80
14	12-13 July 2012	NIRD Hyderabad	Data View Module Verification and Correction Modules and on Claims and Objections Procedures (COTS)	60
15	23-24 July 2012	NIRD Guwahati	-do-	65
16	23-4-2014	NIRD Hyderabad	COTS to BNVs and SIRDs	55
17	28-05-2014	-do-	-do-	28
18	State Specific trainings on V& and COTS was held in the following States in 2013: Tamil Nadu, Pondicherry Kerala, Karnataka, Orissa, Chhattisgarh, Chandigarh, Uttarakhand, Bihar, Uttar Pradesh, Jharkhand, Maharashtra and New Delhi			13 x 60= 780 approximate = 780
19	BNVs trained at SIRDs at Tamil Nadu, Kerala, Maharashtra, Andhra Pradesh, Madhya Pradesh, Uttar Pradesh, Gujarat and Rajasthan (in 2014)			8 x 100= 800
20	Total			2635

residents/households within its jurisdiction. Observations of the Gram Sabha regarding inaccuracies in the information displayed will be recorded. The recording should be made in the prescribed Form E and would be part of the resolution of the Gram Sabha and would be treated as Claims and Objections. An individual or household may separately raise claims and objections on the draft publication by filling the prescribed Form A, B, C and D as applicable, and submit the same to the authorised receiving officer/concerned persons. A date, time and place for the hearing and disposal of the claims and objections will also be announced in the Gram Sabha. The decisions taken at the time of disposal of claims and objections will be entered in the form F. A copy of the filled up form F will also be given to the beneficiary if need be.

The role of Gram Panchayats/PRIs in the conduct of SECC 2011 and tips for the effective conduct of Grama Sabha for settlement of Claims and Objections was elaborated and communicated to all stakeholders. However, due to the peculiarities of the PRIs in the urban areas, it was decided not to go through the public scrutiny in the urban wards sabhas. The Claims and Objections process in urban areas have two stages, viz.: (i) public displaying of Draft List and (ii) claims disposal.

Conclusion

The modus operandi of SECC has come out with valuable outcomes that highlight how the differentiated layers of involvement, coordination and collective nature of tasks can be successfully managed and achieved by multiple stakeholders from public and private sectors in a nation-wide exercise. The efforts to recapitulate and use existing data base from decadal Population Census 2011 as the basic building block and the readiness to use the census concepts with appropriate modifications in SECC has proved to be very effective. The readiness to work together gave both the Census and the Rural and Urban Nodal Departments to identify, reinforce and rectify the strengths and weakness of each other's approaches and systems of data collection. The willingness to break their water-tight existential approach and go hand in hand has set in good examples of successful convergence. The experiment of involving the CPSUs as technology and service providers and who in turn recouped the private sector enterprises from the information technology domain for operationalising the e-modules of data collection via data entry operators/data base administrators and its management, than fully relying on the governments departmental structures and functionaries is a new approach at data collection. This lead to better quality output at lower costs, many a questions likely to be asked in the future as the resources used for conduct of census are from public exchequer. In operationalising the innovative approaches arrived in fixing and estimating various facets of multidimensional poverty, the SECC has made valuable methodical contributions and empirical outcomes. The mammoth, multifaceted and multilayered training and use of human capacity building modules, manpower mobilisation approaches and knowledge dissemination accomplished in SECC have no parallels in the past in the country and in the world. All these efforts have been done in digital/paperless mode and soft data is made available for scrutiny, use, updation and made it possible for integration with those from different sources (e.g. Aadhaar), and in execution of various schemes (e.g.MGNREGS, IAY,NSAP etc) shall help achieve critical role of identification of beneficiaries and proper targeting of scheme benefits that leads to faster development of the nation.

2

SUDHIR KUMAR TEWARI and N.K. SAHU

SECC in Urban Areas[1]

In rural areas, the methodology to identify beneficiaries for determining their eligibility to avail benefits admissible under relevant schemes has been in existence for quite some time. It includes the country wide BPL (below poverty line) census conducted by Ministry of Rural Development in 1992, 1997 and 2002 at the beginning of respective Five Year Plans. No similar exercise, however, was ever undertaken in urban areas. Therefore, unlike rural areas, no uniform methodology has evolved for identification of poor households in urban areas.

In the absence of a uniform national methodology, respective State government/UT administration have adopted different approaches to conducting BPL survey e.g., assessment of household income based on self-reporting and personal verification by local officials, which lead to highly subjective and unverifiable outcomes and result in incoherent poverty estimates across the country. Accordingly, since there is lack of transparency, there are allegations of large exclusion errors and diversion of resources in implementation of various programmes for urban poor.

The erstwhile Planning Commission had constituted an Expert Group in May 2010 to recommend detailed methodology for identification of BPL families in urban areas. The Expert Group submitted its Interim Report in May 2011 and recommended that urban BPL survey be conducted concurrently with forthcoming round of BPL survey in rural areas, as per questionnaire format for urban areas, which formed part of the Interim Report. Government

1. The comments/observations are personal and do not reflect the position of government where the authors work.

approved the approach recommended in Interim Report of Expert Group for implementation as part of combined Socio-economic and Caste Census (SECC), which commenced in June 2011. However, the final report came in 2012 which identified the main features of urban poverty as informalisation of labour markets, migration patterns, poor living conditions, problem of housing, lack of access to education, lack of access to healthcare, lack of access to financial services, and lack of social safety nets.

Accordingly, the expert group suggested a threefold classification for identification of urban households on the basis of: (i) residential vulnerability (households living in houses of more than one room with roof of grass/thatch/bamboo/wood/mud etc. and wall of grass/thatch/bamboo etc.; households living in houses with roof of handmade tiles or g.i./metal/asbestos sheets; and wall of mud/unburnt brick or wood or stone not packed with mortar or g.i./metal/asbestos sheets, household with non-availability of drinking water source within or near the premises, households with main source of lighting other than electricity and households with no exclusive water-seal latrines), (ii) social vulnerability (female-headed households i.e. households where there is no adult male member or where the principal bread-earner in the family is a woman; household with a widow below the age of 50 years as a member of that household; SC (scheduled caste) households; ST (scheduled tribe) households; no literate adult; no adult in the household educated up to primary level and disabled/chronically ill person in the household, and (iii) occupational vulnerability (occupation of the head of the household street vendor/cobbler/hawker, construction/plumber/mason/labour/painter/welder/security guard, home-based/artisans/tailor, transport worker/driver/conductor/helper to drivers and conductors/ cart puller/rickshaw, washermen/*dhobi*/*chowkidar*, coolie/head-loader, shop-worker/assistant/helper/peon in small establishment/attendant/waiter, electrician/mechanic/assembler/repair worker; households where the main source of income of the head of the household is through a weekly/daily wage earning and households where there is no enterprise/wage earning i.e. non-work and the earning is chiefly through any one of the means i.e. pension; interest; and/or rent).

The objective in this chapter is to draw attention to some specific aspects of the urban landscape, which are distinct from rural hinterland, and accordingly to highlight the tailor-made approaches, which were adopted to conduct the SECC in urban areas.

Governance Structure

'Gram Panchayat' and 'Gram Sabha' are age old institutions that exist uniformly in all rural areas across the length and breadth of the country. They have robust governance structures, which are involved with disbursing benefits under various schemes of rural development. In urban areas, the Constitution of India provides for a Municipal area to be divided into 'wards' and also formation of Wards Committees, consisting of one or more wards, within the territorial area of a Municipality. The Legislature of a State may, by law, however, make provision with respect to the composition and territorial area of a Wards Committee and the manner in which seats in a Wards Committee shall be filled.

Local government is in the State list and respective State legislature may assign functions to ULBs (urban local bodies). The legal, institutional and financial arrangements regarding assignment of functions to municipalities vary across States. The conformity legislation enacted by respective State government in the wake of the Constitution (74th) Amendment Act, 1992, is itself not uniform. The Government of India has a limited role as regards devolution of more powers to municipalities and/or promoting a particular model of decentralisation.

Densely Located Households

As compared to traditional rural setting, which is by and large dispersed, urban landscapes represent conglomerations which are primarily amorphous. Delineating and demarcating the densely located households within, which comprise the target of SECC, constituted the initial challenge in commencing the exercise in the field. The magnitude and dimensions of the challenge may be highlighted with reference to few prime locations. National Capital Territory of Delhi with population of 17.8 million (2011) and geographical area of 1,483 km^2 yields population density of 11,320 person per km^2, which is the highest among 36 states/union territories in the nation. The concentration of population within various states, however, may be more acute e.g., with reference to the population, as per Census of India 2011, the density of population in metropolis of Chennai, Municipal Corporation of Greater Mumbai (MCGM) and Kolkata megacity would work out to 16,638 per km^2, 20,482 per km^2, and 24,000 km^2 respectively.

Slums

It is not only density of population, but also the complex nature and form of dwelling units, per-se, in specific situations that pose a formidable problem in enumeration. Slums, as a house type constitute the single largest group in this regard. Concerted focus on their enumeration, however, is of recent origin. The Census of India 2001 presented slum demography on the basis of actual count during population enumeration. A Committee on Slum Statistics/Census (2008) constituted by Government of India offered insight into conceptual and methodological issues associated with classification and enumeration of slums. The Census of India 2011 released datasets on housing stock, amenities and assets regarding slums based on the house listing and housing Census for the first time.

As per Census of India 2011, 13.9 million households comprising 65.49 million persons were enumerated as living in slums in 2,613 slum reporting cities/towns. The slum population represented 22.4 per cent of the total population of slum reporting cities/towns and 17.4 per cent of total urban population of all States and UTs. Cities which reported high proportion of population residing in slums include Jabalpur (45%), Greater Vishakhapatnam Municipal Corporation (44%), Meerut (42%), Greater Mumbai (42%) and Raipur (39%).

Administrative Boundaries

The administrative boundaries, associated framework and methods followed during the conduct of Census of India 2011 were adopted to carry out the SECC. This ready template proved to be of great value in urban areas, especially since urban space is densely populated with various categories of buildings like offices, commercial establishments, production and service centres, residential dwelling units etc. The AHL (abridged house list) with building/census house numbers and layout maps came handy for conducting the SECC. During the SECC exercise, however, relevant particulars regarding actual buildings and census houses were updated and the same were captured on the AHL and layout maps. For updating/renumbering, the field staff was instructed to follow the relevant guidelines prescribed by the Registrar General & Census Commissioner, India. Two sample layout maps pertaining to NCT of Delhi are given in Annexure A-2.1 and A-2.2 for illustration. The 1st layout map contains 291 AHL households that have 226 images and 65 houses without images. However, 291 total households were enumerated in this EB (enumeration block). In 2nd map, though there were no images of 433 houses, enumerations were done for 448 households. Similarly, the EBs carved out during the population enumeration phase of Census 2011 were used in SECC, which provided a firm basis and introduced a systematic approach for carrying out the entire exercise.

Transparency, Accountability and Public Participation

Transparency, accountability and public participation was prescribed at each stage of SECC, commencing with canvassing of questionnaire at enumeration stage to organising the public scrutiny protocols towards the concluding phase. The necessary processes in this regard designed to achieve the outcome are described in chapter 3 ahead. In rural areas, public scrutiny of data was done at three stages. In urban areas, however, due to peculiarities associated with the functioning of ULBs, two stage public scrutiny was envisaged. It was provided that according to the prevailing local conditions, household particulars including socioeconomic data may be displayed in prominent places e.g., ward office, designated place where claims & objections forms are to be received, municipal corporation/town municipal office, community halls in areas predominantly inhabited by urban poor/slum-dwellers etc. Accordingly, claims & objections were invited, received and disposed off following specific urban protocols.

There was no provision for calling meeting of Ward Sabha, a la Gram Sabha, and the SECC data in urban areas was not scrutinised publicly. Accordingly, Format E was not prescribed in urban COTS (claims and objections tracking system). By using the remaining forms, however, respondents in urban areas could raise objections on collected data for effecting desired modifications and deletions/ additions, as per procedure.

Capturing the Location Address of Households

The questionnaire adopted for SECC did not have provision to capture the geographical address of the household enumerated in both rural and urban areas. During preparation of NPR, however, the postal address and postal pin code had been captured. Therefore, to facilitate the process of identification of household/individual, the address particulars captured during the NPR exercise were migrated and merged with SECC database. Such integration of data was performed by NIC by mapping the common AHL TIN in both the NPR and SECC data bases. Preliminary results suggest that this mapping may be feasible for more than 75 per cent of the total population covered in the combined SECC 2011 data base.

Constraints and the Way Forward

As mentioned above, the questionnaire and methodology for conducting SECC in urban areas was prescribed in the interim report of an Expert Group constituted by erstwhile Planning Commission. The interim report of the Expert Group was submitted in May 2011. Government decided to launch the combined SECC-2011 in June 2011 with directions to complete the exercise by December 2011. The questionnaire prepared for urban areas, however, was neither tested in the field nor was a pilot survey conducted prior to canvassing the entire schedule for SECC-2011.

Combined SECC-2011 project was originally envisaged to be completed in three major phases comprising enumeration, supervision and publication followed by claims & objections. At an intermediate stage of the exercise, however, it was considered necessary to introduce a V&C (verification & correction) module primarily to correct for incomplete/erroneous returns of caste/tribe name and other discrepancies observed in the data regarding PTGs (primitive tribal groups), MSs (manual scavengers), LRBL (legally released bonded labour) and occupational categories viz., beggars and rag pickers. Accordingly, field functionaries in urban areas were advised to verify and correct the relevant discrepancies and ensure accuracy of data.

The final report of the Expert Group was submitted in December 2012. It recommended a 3-stage methodology comprising: (i) Automatic exclusion of relatively wealthier households, (ii) Automatic Inclusion of more vulnerable households, and (iii) Scoring index to determine households which would fall in between automatically excluded and automatically included, on the basis of specific criteria/ parameters to be used at each stage. By then, the combined SECC-2011 project, however, had already been launched in all States/UTs, as per specific approval of Government and it was in progress. Also, for various reasons, which are common to both urban and rural areas and have been discussed elsewhere in the publication, the project suffered significant time and cost overrun and could be completed only by March 2016.

The SECC raw data for urban areas, as per questionnaire canvassed, was placed in public domain (*secc.gov.in*). It was decided that since SECC constitutes the largest and most comprehensive latest available nation-wide database of households and their socioeconomic characteristics, it would be appropriate that it is utilised by different departments to target beneficiaries for various programmes. Individual welfare programmes of various departments, however, will have their own respective targeting criteria based on SECC data. Guidelines of Pradhan Mantri Awas Yojana (PMAY)/ Housing for All (HFA) (Urban) Mission, which is administered by Ministry of Housing & Urban Poverty Alleviation, provide that under the 'beneficiary-led individual house construction or enhancement' component, the ULBs shall validate the information given by the beneficiaries and that the condition of the house of the prospective beneficiaries should be checked with SECC data so as to ensure beneficiary's consequent eligibility for construction of new housing.

Completion of SECC-2011 in urban areas and construction of socioeconomic data base at an all-India level mark the beginning of efforts to identify eligible prospective beneficiaries for various purposes. Issues regarding periodic updation of this data base, its actual use across the board, convergence with other platforms and portals etc. constitute some of the challenges which lie ahead and would require effective resolution to derive maximum advantage from the project and realise the outcome.

Annexure A-2.1

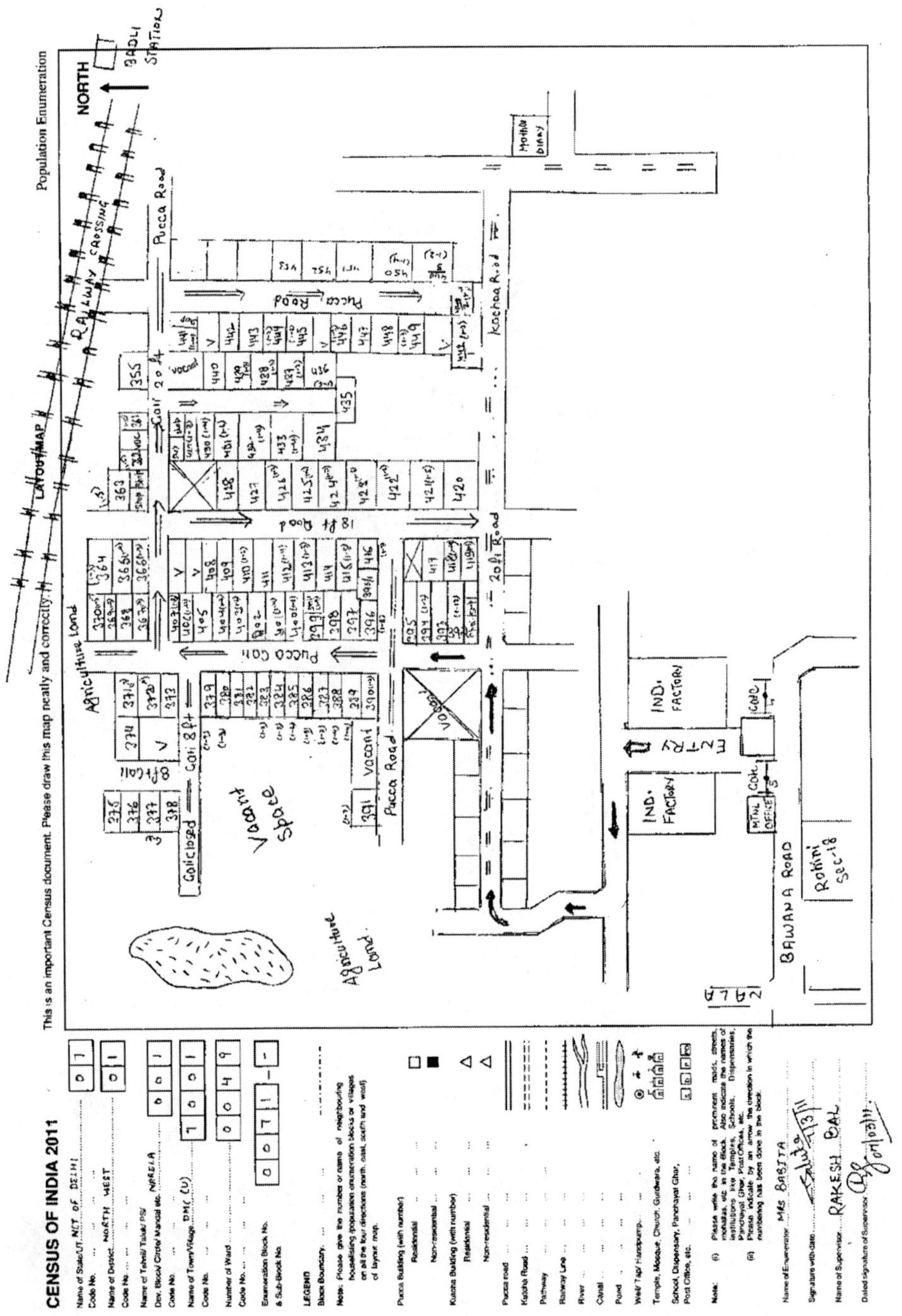

Annexure A-2.2

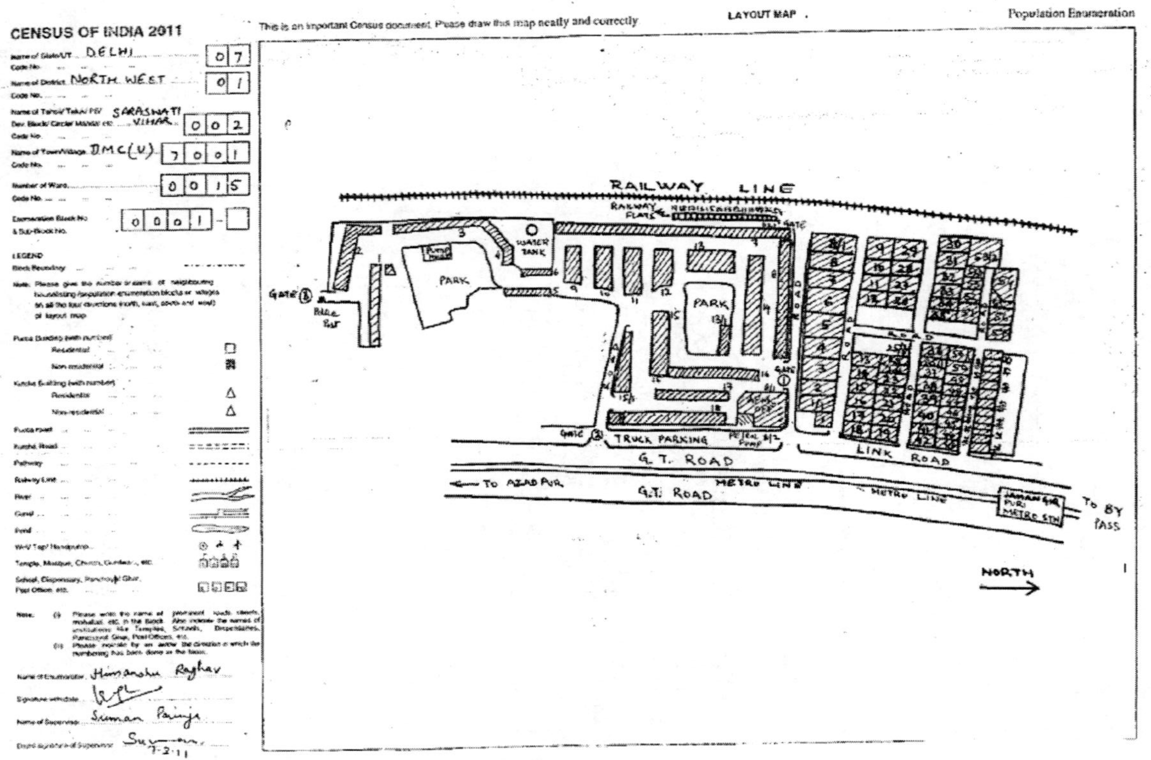
CENSUS OF INDIA 2011

This is an important Census document. Please draw this map neatly and correctly

LAYOUT MAP

Population Enumeration

Name of State/UT DELHI 07
Code No.
Name of District NORTH WEST 01
Code No.
Name of Tahsil/Taluk/PS/Dev. Block/Circle/Mandal etc. SARASWATI VIHAR 002
Code No.
Name of Town/Village DMC(U) 7001
Code No.
Number of Ward 0015
Code No.
Enumeration Block No. & Sub-Block No. 0001 -

LEGEND
Block Boundary
Note: Please give the number or name of neighbouring houselisting/population enumeration blocks or villages on all the four directions (north, east, south and west) of layout map

Pucca Building (with number)
Residential
Non-residential
Kutcha Building (with number)
Residential
Non-residential
Pucca Road
Kutcha Road
Pathway
Railway Line
River
Canal
Pond
Well/ Tap/ Handpump
Temple, Mosque, Church, Gurdwara, etc.
School, Dispensary, Panchayat Ghar, Post Office, etc.

Note: (i) Please write the name of prominent roads, streets, mohallas, etc. in the Block. Also indicate the names of institutions like Temples, Schools, Dispensaries, Panchayat Ghar, Post Offices, etc.
(ii) Please indicate by an arrow the direction in which the numbering has been done in the block.

Name of Enumerator Himanshu Raghav
Signature with date
Name of Supervisor Suman Painja
Dated signature of Supervisor 7-2-11

3

PURNENDU K. BANERJEE and JOSEPH ABRAHAM

Caste Data Collection in SECC[1]

Introduction

In India, the only Census where data on caste name was collected and published was in 1931. This exercise was repeated in 1941, but data was not tabulated/published, as the Second World War began by that time. The demand for a caste census along with population census in 2011 was raised on the floor of the Indian Parliament. By that time, the Census questionnaire was already finalised. Therefore, after a detailed examination of the demand by a Group of Ministers Chaired by the then finance minister (Shri Pranab Mukherjee, who afterwards became the President of India), it was decided that the caste name related data shall be collected along with the BPL Census. The debate on caste census has come to recent headlines in the context of the 2011 census. More intense debate on caste related issues had been going on in academic and political circles over the past few years. The OBCs (other backward castes) and Dalits' political movements and their arguments for conduct of caste census maintain and strengthen the sensitivity and reflect the aspirations within their respective communities. In the past different kinds of census had brought to light different kinds of social discriminations, and thus constituted various forms of democratic social knowledge. Some have argued that the caste census is resisted by the upper castes for fear of being exposed in their disproportionate control of resources. There seems to be constitutional provisions for recognition of backward classes and the same used as a basis and arguments raised for a census that counts the OBCs. The current debates on caste census

1. The comments/observations are personal and do not reflect the position of government or organisations where the authors work/worked.

has helped to bring to front two important issues; one, the effective distribution of resources to people in proportion to their numbers, which should be freed from the grip of dominant elite in the society and two, the democratic potential of the caste census in the long march towards the annihilation of caste, should be kept firmly in view. In recent past (winter session, i.e., November 2009), the Lok Sabha—the lower house in Indian Parliament—witnessed debates and discussions on the need for conducting of caste census. Given this brief historical background, it is proposed to describe below the present status, salient elements, problems and processes of the conduct of caste data collection as an integral part of Socio-economic and Caste Census (SECC) 2011, which is perhaps world's first modern Census, that too done by India.

Caste-based Census: Prime Minister Assures Lok Sabha

Prime Minister of India on 6 May 2010 assured the Lok Sabha that the government would soon take a decision on the opposition demand for a caste-based census. "I am aware of the views of the Members of Parliament belonging to all sections. I assure you that the Cabinet will take a decision shortly," Dr Singh said in a brief statement, the statement put at ease agitated opposition members, particularly Samajwadi Party chief Mulayam Singh Yadav, Rashtriya Janata Dal supremo Lalu Prasad and Janata Dal (United) leader Sharad Yadav. These three leaders and other opposition members were ruffled over Home Minister P. Chidambaram's reply to the debate on census. He said there were difficulties in conducting a caste-based census and the government would keep in mind the members' views. As soon as the Minister concluded his reply, which was punctuated by interruptions from the opposition benches, and the House was adjourned to meet again at 2.30 p.m., and then these three leaders approached the Prime Minister, Congress president Sonia Gandhi and leader of the House Pranab Mukherjee. They were seen talking to them, trying to elicit some kind of an assurance on their demand. They also held a meeting with Mr Mukherjee separately, where BJP's deputy leader Gopinath Munde was also present. When Dr Singh gave the assurance soon after the House reassembled, the three leaders thanked him and the Congress president for trying to resolve the matter in an amicable manner. Earlier, while taking note of the discussion over the last two days in Lok Sabha that witnessed arguments and reasons for canvassing the question of caste, Home Minister Mr Chidambaram said the government was seized of the matter. "The views of honourable members will certainly be a valuable guide to the government," he said in his statement, which virtually gave out nothing but offered explanation relating to the Census 2011 and preparation of the National Population Register (NPR), the two exercises which were on. The debate on the issue saw members cutting across party lines favouring a caste-based census. Replying to the debate, the Home Minister stressed that the main aim of the exercise was headcount and listed difficulties in including caste in the list. The Registrar-General had pointed out the existence of a number of logistical and practical difficulties in canvassing the question of caste while conducting the census. However, he assured the House that the government "will give due weight to all aspects of the issue." "The enumerator is not an investigator or verifier," Mr Chidambaram said, pointing out that 21 lakh enumerators—mostly primary school teachers—had no training or expertise to classify the answer as OBC or otherwise."There is a Central list of OBCs and State-specific lists of OBCs. Some States do not have a list of OBCs while some have a list of OBCs and a sub-set called Most Backward Castes," he said.

Many Issues

Quoting the Registrar-General of India, Mr Chidambaram, Home Minister said issues regarding methodology, avoiding phonetic and spelling errors, stage of canvassing of caste, maintaining integrity of enumeration and doing an accurate headcount of population etc. would arise. He said:

> The census is done under the authority of the Census Act, 1948. Census 2011 will be the 15th national census since 1872 and the 7th since Independence. Population census is the total process of collecting demographic, economic and social data. What is published as the census data are only aggregates; the information relating to the individual is confidential and not shared with anyone or any authority.

While throwing up the question, if the census was the vehicle to carry out caste-based enumeration, he pointed out that records showed that an attempt was made by the Ministry of Social Justice to include caste as one of the questions that should be canvassed in the 2001 Census. "However, the government of the day—the National Democratic Alliance (NDA) government—did not take a decision to that effect and maintained the policy that has been in force since 1951".

The current fact of the matter in the country is that even after having progressed on so far ahead, caste still plays a vital and decisive role not just in domestic and social arenas, but in political affairs too. The emphasis on caste on such a large scale, like in the census, needs to be discussed more seriously. The number of castes that have been listed under OBC/SC/ST categories need to be updated. If government is still going ahead with quotas, they need to keep a record of the progress made by those castes that come under reserved categories. It is advisable that a better way be found to improve the lot of those who fall into the different caste categories. Also it is argued on the contrary that, in this 21st century, instead of having a census based on caste, more attention should be given to education, health, technological advancement, sanitation and other welfare activities, so that our dream of becoming a developed nation doesn't get blurred.

Caste Census in SECC

The structure of SECC 2011 was developed on the strong foundation pillars built up on by the Office of the Registrar General and the Census Commissioner (ORGI) of Government of India (GoI) in the conduct of regular decennial population Census in the country. An array of the strong building blocks used in SECC was provided by the ORGI. A few of these being the National Population Register (NPR) and House listing data base, EBs (enumeration blocks), AHL (abridged house lists), house numbering, layout maps, the concepts and definitions used and in streamlining conduct of the Caste Census related tasks and to prepare socioeconomic profile of various castes etc. The valuable and strong support of the DCOs (directors of census operations) of each of the State/UTs had been another noteworthy input from the ORGI for smooth conduct of SECC.

Recording Caste Returns

The SECC was designed to capture name of caste/tribe for all persons of the country, including the SCs (scheduled castes) and STs (scheduled tribes). As the time gap between the population census and SECC was not much, any large variation in the counts can be questioned by the users and the concerned source agencies would have to provide a great deal of explanation for that. Moreover, the population census releases population counts of each village and each ward of the country. The SECC also released draft lists of person level data at each area. Building the backbone of such a transparent exercise, for more than 1.2 billion persons of the country was a major challenge for all the concerned Ministries, particularly the ORGI, who were providing the technical support for this exercise.

The population Census questionnaire is canvassed under the Census Act, 1948 and individual information is confidential by law. In 2010, the ORGI conducted data collection exercise for the NPR, where person level data, like name, gender, age, etc. were recorded. The NPR data was also displayed in local areas for correction and a similar exercise was to be followed for the SECC later. The scanned schedules of the NPR were available by the time SECC started. To have some control on the data collection exercise and to avoid duplication, a part of the scanned images were uploaded in the handheld devices and the same was verified and updated during field-work of SECC. In India, such an exercise was never done before. The scale of such an exercise can be understood if one keeps in mind that the SECC final results show a total of 244.92 million households in India, which is only marginally lower than the total households reported in the decennial population Census (249.5 million). In SECC, after checking details like name, gender, and relationship to head of the household, fresh data was recorded for month and year of birth, religion, caste/tribe status and caste/tribe name were recorded for each member of the households. Since data collection was in the TPC (tablet personal computer), these details, for the first time, were available in soft form without any further data entry effort. Keeping parity with the instructions followed in population Census, the status of a person was recorded as SC only if the declared religion was one of Hindus, Sikhs or Buddhists. For recording a person's caste/tribe status as ST, no such restriction with respect to religion was there. Name of caste/tribe was recorded as per the declaration of the respondent, respondent being a member of the household. Instructions to the enumerators was simple, respondent's replay is to be faithfully recorded (realising very well that mistakes may arise due to varying levels of knowledge of the enumerators, data entry operators and the medium in which it is recorded in the TPCs). It was decided to identify errors subsequently using appropriate software solutions for either keeping them as it is or for carrying out corrections and harmonisation. Caste/tribe recording based on the respondent's reply only were noted in the SECC-2011 and acknowledgement slip issued by the enumerator to the respondents clearly mentioned that "the caste/tribe is as declared by the respondent and it will not confer any right to claim any caste/tribe/community certificate".

Verification & Correction Module

On completing enumeration and supervision stages of SECC, the V&C (verification & correction) module was introduced. The latter was necessitated as it was noted through a DVM (data view module) that data collected contained several lacunas. These were of the following type: (i) a very high number of households reported to have members belonging to PTGs (primitive tribal groups), LRBL (legally released bonded labourers) and MS (manual scavengers); (ii) in the urban areas, numbers of occupational categories like beggars and rag picker were found to be unreasonably high; and (iii) apparently wrong/incomplete returns of caste/tribe names, e.g. 'other'/'open'/'general'/'OBC'/abbreviated caste name/religion name/ repeated characters like "....", etc. have been found in the enumerated and uploaded data. The NIC carried out the soft-runs on the raw data uploaded in the central server, based on requests of analysis from the concerned Ministries, like the ORGI. Having soft data uploaded in a central server right from the beginning of field-work, in spite of several difficulties made such an exercise possible. Remember that the expansion of mobile networks, etc. did not happen till then and in many places, the enumerators had to travel long distances to get their data in TPCs loaded in desktop computers at *tehsils*/blocks and then sometimes data in hard drives had to be sent for loading from State headquarters. However, this trouble was worth, as within a few weeks of data collection, one could run checks and identify errors before the payments were made to the part-time enumerators. Once the data discrepancies were brought before the other stakeholders, like concerned State governments, many of the states desired to get these verified and corrected before the Draft List is printed and published for scrutiny in the public domain, though caste status would not be revealed in this list.

The V&C Module was introduced in October 2012. This was also a first in such a large scale exercise. The nodal Government Ministry mandated that all the V&C files of a district are processed and uploaded by states before they are permitted to generate SECC Draft List files of that district. It is to be noted that, these SECC Draft List containing details of each household (except religion and caste) are to be displayed in villages and wards. Based on the experience of conducting the V&C exercise, some states have called for issuance of clear instructions as to how corrections in the data on caste names are to be done in a time bound manner.

Caste Names' Error Corrections

The verification and correction module had some errors with respect to the parameters declared for determining the BPL status, like bonded labour, manual scavengers, etc. in rural areas and beggers, etc. in the urban areas. Number of such errors was far less than the number of apparently wrong/incomplete caste names filled in the data. It was found that the pace of uploading of files with updated caste data was lagging behind the uploading of files with revised data of other types of errors, for many states/UTs. In some cases, same files were uploaded without revising anything. For files where caste corrections were done, some of the corrections still persist with inconsistent/unconfirmed caste names. All these prompted ORGI to introduce the additional V&C activity. The additional 'C-files' were permitted for those EBs wherein the corrections needed were above a threshold number. In prior review meetings held, it was observed by States/UTs that the additional caste error corrections are still delaying the publication of Draft List. As a response to the above, it was decided in the PRC meeting held in 2014 that the correction and uploading of additional V&C files would not be insisted for draft list generation and publication. However, it was decided that the processes of additional verification and corrections of caste data can continue till end of the COTS (claims and objections stage), i.e. till the generation of final list (F-Files). Necessary guidelines on the conduct of caste name error corrections were issued by the Ministry of Rural Development (MoRD) in October 2014.

Caste Census Results

As part of SECC, the first countrywide caste census of independent India has been also undertaken, but so far no final progress report is made available by the Ministry of Home Affairs and the Census Commissioner of India as to, when the data collection will be over, when processing be taken up and final list put up in the public domain. However, a committee headed by NITI Ayog Chairman has been set up by Government of India to look into the caste data being collected as part of SECC 2011. Minister of State for Home Hansraj Ahir, on 10 August 2016 said the caste- related data, as enumerated in SECC, has not yet been released. "The government has decided to constitute an expert group under the chairmanship of Arvind Panagariya, Vice-chairman, NITI Aayog to classify and categorise the caste names returned in SECC," he said replying to a written question by honb'le MP Shri Veerendra Kumar. No formal notification is seen to be issued in the subject.

Role of ORGI/DCOs

The States have been asking for assistance from ORGI/DCOs to prepare a list of the correct caste names to facilitate the smooth completion of field verifications. Besides these, the States/UTs have been calling for appropriate guidelines from ORGI so as to carry out the pending caste name error corrections. It is obvious that these have to come in from the ORGI and DCOs concerned and necessary directions provided to DCOs to work in liaison with State SECC nodal officers to help complete the caste data collection exercise. However, no such complete list of all caste names of India exist. One such exercise, called the 'People of India' project was taken up by the Anthropological Survey of India (ASI) during 1985 to 1992 using available data from the Censuses of 1881 to 1941, SC/ST lists, OBC lists, lists used by the Mandal Commission, etc. They identified more than 4,600 'communities' and published the findings in 11 volumes. The fact that even the ASI, after using more than 600 trained scholars for so many years, could not identify all the communities in India, tells about the complexity and vastness of this exercise. Thus, blaming one organisation/stakeholder or the other would essentially undermine the enormity of the task at hand.

Conclusion

The SECC project has come to an end formally in March 2016. Though rural and urban household data that were collected have been published, status of caste data is yet to be finalised and published.

Bibliography

1. Q.16015/04/2011-AI(RD)/GoI/MoRD/Clarifications on Caste Corrections issued dated 13/11/2014 issued by the EA-MoRD, Dept. of Rural Department. New Delhi.
2. XIII Lok Sabha Debates, Session II (Winter Session) Tuesday, 30 November 1999. Matters under Rule-377. Que. by Sri. D. Venugopal (Thirupattur M. P).
3. *The Hindu*, New Delhi, 7 May 2010. "Decision on Caste Based Census, Manmohan Assures Lok Sabha".
4. *Anveshsi* 1(2), June 2011, "Caste, Census, Democracy Different Perspectives". (See editorial and four articles).
5. *Websites*:

   www.https://sec-census.nic.in

   http://secc.gov.in/statewiseSeccDataSummaryReport?reportType=SECC%20Data%20Summary

   https://ansi.gov.in/people-of-india/

4

JOSEPH ABRAHAM

Transparency, Accountability and Public Participation in SECC

Introduction

The legal frameworks within which public sector entities operate are different from that of private. The government entities are statutorily required to conduct their business through open and transparent processes to ensure that they are accountable to the citizenry. This modern practice of open governance is viewed as both a key feature and a necessary condition of a modern democratic state. It is based upon this conviction that the people can effectively exercise their constitutional role as overseers of government action where their unfettered rights of access to information about government operations are secure. Even though the Socio-economic and Caste Census (SECC) did not have any statutory backings, as is with decadal population census done in India with the support of Census Act 1948, it has gone beyond such an act and taken care of the requirements of transparency, accountability and public ownership in a systematic manner. The key features of the above are mass media coverage, use of video clippings, issue of press releases, wall writings, conduct of claims and objections procedures, setting up of toll free call centres and its online and web based tracking in very rigorous ways. By ensuring the above, the whole work of SECC 2011 has been done in a transparent, accountable, acceptable and scientific manner. The evolution and upkeep requirements of various management components of this framework in SECC are illustrated here. The SECC is acclaimed to be world first ever massive paperless digital data collection initiative. It would be interesting to know as to how this has been accomplished in this world's largest democracy in a transparent, accountable and verifiable manner. Given the specific methods and technologies used, if the data collected are erroneous beyond permissible levels, then what else or how to improve upon the experiences gained would be called in very soon. The Ministry has decided to make use of the SECC data for beneficiary selections and for fixing the eligible resource allocations under various rural development programmes. Given the above perspective, let us elaborate how concerns of transparency, accountability and public scrutiny are taken care of in various SECC processes.

Canvassing the Questionnaire

How concerns of transparency and citizen's responsibilities are taken care of in the data collection processes of SECC is elaborated here. In the use of questionnaire, several safeguards were insisted upon.

After entering all the responses to questionnaire particulars, the enumerator is mandated to show the respondent what data they have entered in the HHD (handheld device) (TPC [tablet personal computer]) and confirm from his/her that all particulars got have been entered properly. In case a respondent is illiterate or cannot see the entries, then he reads it out to the respondent and explain the entries. Only after a respondent confirms the details, then the enumerator has to proceed to write out the acknowledgement slip. After the respondent has confirmed the data, the acknowledgement slip duly signed by enumerator is issued to each household. The signature/thumb impression of a respondent is to be obtained on the office copy of the acknowledgment slip. The Enumerator and DEO (data entry operator) will have to jointly declare in the acknowledgement slip the following:

(i) We declare that all the information provided by the respondent has been correctly entered in the HHD and shown, read and verified by the respondent.

(ii) The individual/household data on religion and caste/tribe is confidential and will not be published.

iii) The caste/tribe is as declared by the respondent and does not confer any right to claim any caste/tribe/community certificates.

The acknowledgement slip included the following declarations that were to be signed by a respondent:

(i) I declare that all the information provided to the enumerator is true to best of my knowledge and belief.

Linking Acknowledgement Slip, TPC, and SECC Central Database

The Acknowledgement Slips were issued to respondent households as a mark of completion of data collection as per laid out procedures and terms. The data collected was to be read out to the respondent by Enumerator and consent obtained before closing the session and then the slip is to be issued. The data collected is stored in the Tablet PC and then subsequently transferred to the data charge centre server. Then as a next step the information collected to be uploaded to the main data base server located with the NIC. This schema of links may not have worked meticulously for various reasons. The main failure/gap in the procedures were that the Enumerators have not been explained the terms and conditions to respondent households, besides not issuing the acknowledgement slip. Such error cases resulted in observing a large turnover of non-willingness to place data in public domain as was noted in some States, and had to be corrected through a repeat procedure. Even an advertisement was made in local news papers by each state government before the Claims & Objections stage.

(ii) I have seen the information recorded/the information has been read out to me and I certify that they have been entered correctly.

iii) I am aware that the caste/tribe is as declared by the respondent and does not confer any right to claim any caste/tribe/community certificates.

iv) I am aware that my personal/household level data, except religion and caste/tribe will be published. I give consent to/or I do not give my consent for the same. I am aware that by not giving my consent for publishing my data in the public domain, I will not be considered for benefits under the welfare schemes of the Government.

v) If the respondent refuses to sign or affix the thumb impression please write 'Refused to Sign'.

vi) Only after the contents have been verified by the respondent and the acknowledgement slip has been signed by the respondent, should the data be saved in the HHD. In case of refusal to sign, the Enumerator should note this in the column provided in the HHD and then save the data. All such cases would be referred to the Supervisor.

Important Points to be Kept in Mind during Enumeration

The Enumerator along with DEO discharges major functions that ensure transparency and public accountability. The respondent is responsible to furnish information, but the Enumerator has opportunity to record remarks in case of doubt of authenticity and correctness from the respondent. The following provisions ensure these rights and duties:

(i) Every household in the EB (enumeration block) will have to be visited by the Enumerator and DEO to collect the information;
(ii) No proof or document should be asked form the respondents in confirmation of any information provided by;
(iii) In case the Enumerator feels that any information given by the household is incorrect, s/he shall not argue with the respondent, but can record observations in the space provided for enumerators remarks and notify to the Supervisor;
(iv) Do not collect any information from respondents who are not members of the household;
(v) To the extent possible, collect information from adult members of the household;
(vi) Do not reveal information given by one household to others;
(vii) Ask the respondent all the questions; and
(viii) Always be gender sensitive and do not let personal biases to come in.

Mandatory Declarations

The following concerns are mandated upon the respondent and the Enumerator. At the end of canvassing the information, the Enumerator must ask following specific question to the respondent. 'Is the household giving its consent to place the information in public domain, except data on religion and caste/tribe? Then appropriate code is to be entered in the HHD. Then the Enumerator must record appropriate code (Code-1=Yes and Code-2=No) on whether s/he agrees with the responses given by the respondent. If the answer is 'No', then detailed reasons should be recorded within 100 characters. Such meticulously designed canvassing methods and recording in TPCs ensures privacy safeguards.

Checks and Balances at Multi Level

Several checks and balances are built into the whole processes of SECC to ensure quality of data collected. The concerns raised, mandates associated and outcomes expected are described below:

Concerns	*Mandates*	*Outcomes*
Enumerator is supposedly/likely to be accompanied by members of Gram Panchayats at the time of field survey and holding Gram Sabha's to ensure a fair process. The IEC campaign also aimed at giving wide publicity and people's participation in all the processes of SECC at the field level, in the Gram Panchayat/Gram Sabha levels.	Wide publicity	Transparency
The data collected in the HHD are saved and protected from being manipulated.	Accuracy/security	Privacy/ security
If the enumerator feels that incorrect information is being provided, s/he may enter remarks within 6 hours of collecting data from particular household. The Supervisor will also visit the house where there are differences and verify the information.	Authenticity	Supervisory verification
The enumerator must report locked households and households where caste/tribe names could not be recorded to a higher level, the Supervisor.	Shortcomings/ gaps to be reported to a higher level	Complete coverage
Third party monitoring is also arranged to ensure that the data is being recorded accurately. The independent monitors were to promote people's participation and facilitate transparency at all levels of conduct of SECC.	Independent monitoring by civil society organisations	Exposure and public scrutiny
Within a time frame, after the publication of the draft list, a Gram Sabha will be called and the information got scrutinised for corrective action.	Gram Sabhas to be held mandatorily	Peoples/ verification/ participation

IEC Campaign and Strategy

Various elements of the IEC (information education and communication) strategies are made use of in SECC to ensure awareness and participation of people in the whole exercise. The well streamlined inputs given for the conduct of IEC were to be put to active use by various stakeholders, but in the field level the latter have not paid adequate attention was evident at the different tiers of execution.

Publicity

The approach to publicity involved fixing of specific tasks, objectives and expected outcomes, as described below. It was clearly specified that information and awareness generation on the SECC process must be carried out through newspaper advertisements, billboards, radio spots and television coverage. IEC material was designed and disseminated to various States/UTs for further use in local language versions. The tasks/objectives/outcomes expected of IEC are given below:

Functionary/Tasks	*Objectives*	*Outcomes*
Transparency/accountability/ verifiability/explicability	To ensure correctness of data collected	Obtain quality data
Peoples participation	Ensure public accountability	People's involvement ensured
Enumerators	Data collection	Specific tasks executed
Supervisors	Over-sear the field coverage and ensure full coverage	Specific tasks executed to ensure full coverage
Verification and corrections	Verification Officers to filed visit and correct errors reported	Ensure public and bureaucratic authenticity.
Claims and objections received	Public scrutiny of data	People's participation

It was proposed to conduct intensive campaigns of about two week's duration in the early stage of the Draft Publication and Claims & Objections. The campaign would be in local/state languages. The following instruments of IEC were prepared at Government of India level and made available to states/UTs.

Posters

These posters were to promote awareness and inclusiveness at all stages of SECC. The themes in these six posters give a main message that, "Your information is your right". The following are the six main captions of each poster: (i) The respondent to give information need not be head of household nor the main person. (ii) Head of the household can be women as well. (iii) Give correct and complete information that will help arrive at your socioeconomic status. (iv) The landless have right to provide right information. (v) From the information you give, your rights will be established. (vi) The differently able also have special rights. These set of six posters were focusing on the need to provide correct information, to facilitate participation of women, and reporting various forms of disability and deprivations etc. The soft copy of these set of posters was provided to states to use it in their local languages to give wide publicity in the SECC campaigns mandated upon.

Video Spots for TV

A 30 seconds video spot was made and circulated to stakeholders. It highlighted how the people can voluntarily participate in SECC and the use of TPCs. The video spot was made in March 2012. It encompassed screenshots encouraging involvement of different sections of the society in typical rural village settings; namely the disabled, tribes engaged in traditional occupations, typical housewife and introduced use of tablet pocket computer to the common man, all belonging to different regions of the country. Thus a pan country rural scenario was made available through introduction of video spots. This has helped the conduct of SECC in a transparent manner.

Radio Spots

Three 50 seconds each duration audio spots were prepared with the help of the National Film Development Corporation (NFDC) and made available to all State/UT governments. These were to be used by the nodal departments for massive awareness campaign on the various features and importance of SECC. These audio spots for broadcasting were designed to explain to general public on procedures that would be followed during the Claims and Objection stage and what is required from them. The objective is to ensure peoples' participation to improve the quality and authenticity of data obtained. The first slot highlighted that all relevant household data that are collected is displayed as 'draft list' at prominent locations of the Gram Panchayat and in the website *www.secc.gov.in* for physical verification and that there is provision to correct errors and to add missing information, by submitting petitions. The use of toll free number to get assistance is also popularised in all these three radio messages. The second slot emphasised the conduct of Claims and Objections procedures and use of relevant forms for effecting necessary changes at the Gram Panchayat level. The decisions of the Gram/Block Panchayat officials on the claims and objections can be put to appeal at the level of district collector and got resolved. And the third slot informed that the collected information will be scrutinised in the special Gram Sabhas held and all the grievances to be got resolved. All the States/UTs were requested to give wide use of these radio spots in the public space and give extensive coverage among the people.

Toll Free Call Centers

A toll free call number has been set up with the technical support of BSNL (Enterprises Business), Faridabad. The select personnel of the BSNL were trained by main stakeholders (MoRD, MoHUPA, Office of the Registrar General & Census Commissioner, India [ORGI] and CPSUs) for the purpose. Regional centres have been set up were in 15 local languages: Telugu, Kannada, Malayalam, Tamil, Assamees, Nagamese, Manipuri, Bengali, Odiya, Punjabi, Urdu, Gujarati, Konkani, English and Hindi; based assistance has been provided to the respondents.

Panchayati Raj Institutions (PRIs) and Gram Sabhas

The PRIs have been involved in the conduct of SECC 2011. The draft SECC List will have to be prominently displayed in the Gram Panchayats—the village democratic administrative unit. The Gram Sabha will have to be called up by *Panchayats* for scrutiny of draft list.

The State/UTs have been requested to involve the PRIs in the IEC campaign. Urban areas (corporations/municipalities/councils) were requested to conduct the same exercise in absence of a counterpart to the Gram Sabhas like in rural areas, but in the urban departments such exercise could not be conducted.

As per progress reports available, about 12.2 crore and 1.6 lakh households respectively from rural and urban areas have raised various claims and objections following due procedures by filing up various types of forms (e.g. A, B, C, D and E) with active participation and involvement of PRIs. About a total of 11 lakh E Forms were filled up in Gram Sabhas. These outcomes indicate the very useful purpose served via the COTS (claims and objection tracking system) procedures prescribed and implemented. The data of this sort is available in the restrictive website *www.sec-census.nic.in*. But this website could only be monitored by those permitted by MoRD, however, the website *www.secc.gov.in* have been made for public access. The procedures prescribed for conduct of Gram Sabhas are the following.

- Within specific days after draft list being published, a Gram Sabha meeting will have to be called for public scrutiny. For this meeting, a representative of District Magistrate/Collector shall remain present. The Gram Sabha shall scrutinise the Draft List for all the residents/households within its jurisdiction. Observations of the Gram Sabha regarding inaccuracies in the information displayed will be recorded.
- The recording should be made in the prescribed Form E which would be part of the resolution of the Gram Sabha and would be treated as Claims and Objections.
- An individual or household may separately raise claims and objections on the draft publication by filling the prescribed Form A, B, C and D as applicable, and submit the same to the authorised receiving officer/concerned persons.
- A date, time and place for the hearing and disposal of the claims and objections will also be announced in the Gram Sabha.

- The decisions taken at the time of disposal of claims and objections will be entered in the form F. A copy of the filled up Form F will also be given to the beneficiary if need be.

The COTS is also developed as a web based online monitoring mechanism for keeping instant track of summary progress of the core outcomes of conduct of Claims and Objections under SECC.

The role of Gram Panchayats/PRIs in the conduct of SECC 2011 and tips for the effective conduct of Gram Sabha for settlement of Claims and Objections was elaborated and communicated to all stakeholders. The placing and scrutiny of draft list in the Gram Sabha if carried out with zeal and enthusiasm by the people's representatives and civil society, in a participatory mode; it would add purpose and bring in data quality. The capacity building of functionaries at grass roots level is the need of time to bring about good governance at all levels of information collection and dissemination by the governments.

Cells for Disposal of Complaints/Grievances

These centres were set up in order to redress the grievances/complaints that are received from public and various other stakeholders. To what extend these cells have been to active use and got positive outcomes is yet to be investigated. However, most of these complaints were of the following issues: complaints of general public and people's representatives on non-coverage, inadequate facilities and low remunerations for the functionaries involved. To address these 'grievance cells-cum-control rooms' were to be set up in the offices of Charge Officers/Principal Census Officer/District Collector/Nodal Officer. These cells were to maintain proper registers and action taken reports to be regularly updated by the responsible officer at each level.

Roles and Responsibilities of Stakeholders

The specific roles and responsibilities of Enumerator/Supervisor/Charge Officer/ Sub Divisional Officers /District/State Principal SECC Officer and that of the functionaries manned by the three CPSUs (central public sector units) were clearly laid down for achieving transparency, accountability and people participation.

Filling, Receiving and Disposing of Claims and Objections: Tracking through the COTS

The COTS refer to the set of activities that are carried out as part of SECC to ensure transparency, and accountability in the public domain. The entire COTS processes and formats were carried out in digital mode. This is a web based system that operates via software's developed by Bharat Electronics Limited (BEL) and supported by NIC New Delhi. COTS have broadly two steps of field level functions to be carried out. But in urban COTS, the format E is not used as Ward Sabhas were not generally held by the urban departments as part of SECC. To receive various Claims and Objections from the respondents and general public on the SECC Draft List published in the public domain, a number of forms were duly filled and submitted. The prescribed forms are for raising objections on collected data, for effecting modifications, and deletions/and additions. These are forms A, B, and C. The form D is for issuing notice to the objector and respondents involved. The form E is for recording the proceedings of Gram Sabhas held on e SECC draft list and F form is for making the corrections in the main database and for issue to respondents. The government officers have to receive, verify and execute corrections as per guidelines as designated disposal officer at Block/*Tehsil*/ Sub-District levels. An appellate officer is also appointed at district level to pass orders on the appellate petitions that will come up from applicants. A timeline of 82 days has been laid out for the conduct of prescribed set of COTS activities in urban and rural areas. Keeping in view local factors, some States had gone for shorter time durations. These broadly include the following set of activities; draft list publication, holding Gram Sabha, receipt of claims and objections, last date for disposal of claims and objections, receipt of appeal petitions, last date for disposal of all appellate petitions, all corrections to be uploaded and prescribed completion certificate to be submitted by States/UTs to MoRD/NIC, and finalisation for printing and publishing.

The BEL was advised to prepare software for use as a server version called the COTS (see Table 4.1). At the DCC (data charge centre), all the details like number and type of applications received and decisions taken on each will be digitised and uploaded to the central server through a system of login IDs assigned to the CPSUs, through the monitoring website developed and managed by NIC. The infrastructure for conduct of online and software based COTS will have to be provided by the states and technical support and manpower services by the CPSUs. The sequential stages of activities to be carried by various 'operators' engaged in the process are given in Table 4.1. These 'operators' are personnel responsible and designated by BEL's venders for online data entry of the COTS processes as done in the field level.

Table 4.1

Structure of COTS Process

Operator-1	*Operator-2*	*Operator-3*	*Operator-4*
C/O forms (A,B,C,E,F)	Operate log in name of designation officer	Do the necessary corrections as recommended	Registering appeals against designated officers decisions
Forms Receiving Officer	Giving decisions (Whether the form is acceptable or not)	Issue filled up F Forms to applicants	Issue decisions of the Appellate Officer
Generate Form-F. Attach with C/O Forms	Forward to next level	Forward information to next level	Upload the final corrections to website/online
Forward for further processing, online web based	Online web based	Online web based	Online web based

Ideally the above described two sets of activities should have been done simultaneously, but practically it has not happened due to field level problems. Shortage of personnel deployed at Block/Tehsil/Gram Panchayat levels, inadequate filed publicity and awareness among the public, lack of IT infrastructure all contributed to the dislocated conduct of Claims and Objections processes in SECC.

Public Notice Prior to Draft List Release

On a preliminary sample analysis, it was revealed in few States that the data made available indicated that about 80 to 90 per cent did opt for not placing the data in the public domain. Whereas the acknowledgement slip has the following declarations that were signed by these respondents, "I am aware that my personal/household level data, except religion and caste/tribe will be published. I give consent to/I do not give my consent for the same (strike off whichever is not applicable). I am aware that by not giving my consent for publishing my data in the public domain, I will not be considered for benefits under the welfare schemes of the Government". This means, the data of these households would not be displayed in public domain for public scrutiny and thus they would be logically out of consideration zone for government scheme benefits. This is a clear case as to how the functionaries do not take adequate care to canvass the questionnaire and failed to disseminate important information to the respondents. In order to avoid such misguided outcomes, a correction procedure was introduced that was not there in the originally designed SECC process.

At the time of data collection, the respondents were to have been briefed of their consent or otherwise of placing the data collected in the public domain except that of the caste name and religion. These inputs on consent are captured in the TPCs. However, it was found through field monitoring and electronic returns that the process of capturing 'consent' of respondents was not foolproof. Thus to be sure doubly, the States/UTs were requested that a public notice be issued in the media well in advance that the SECC data would be put in public domain unless the households submit a written request in the prescribed format that they do not wish to share their data in public. A list of such households will be separately appended in the draft list of each enumeration blocks. If they do not submit such a written request within 15 days of the issue of media notice, then it would have been presumed that there is no objection to share their data in the public domain and proceeded accordingly (for sample notice see Figure 4.1.

Draft List Publication

The SECC draft lists are to be published in the local language. This shall contain basic summary information on the individuals and households and it will be displayed in the public domain for undergoing the Claims and Objections procedures. A copy of the draft list will be made available for display in the office of Gram Panchayat/ BDO (block development officer)/*tehsil* Office and in a prominent place in the *Panchayat* for ready reference by the public, besides web enabling the same in the *http//:secc.gov.in*.

Concurrent Evaluation

Nine IMA (independent monitoring agencies) were enlisted to conduct concurrent monitoring and evaluation of SECC. Besides other, answers to following select six questions are analysed and presented. To each of the questions (Que-1 to 3), the respondents were to give answer options as excellent, very good, good, fair and poor. Reply to Que-4 is in terms of percentages and Que-5 and Que-6 are on various impressions/replays to these questions. It is indicative only and the objective here is to show that there was mechanism to achieve accountability in the SECC processes. The six main questions canvassed by IMAs are: Question-1: data/ information display in villages/GPs; Question-2: awareness levels of local population regarding the draft list and its publication in GP; Question-3: publicity on holding Gram Sabha meeting; Question-4: whether Tablet PCs were used by the Enumerators/Supervisors/ Data Entry Operators to know to what extent these Tablet PCs were used; Question-5: whether both Enumerators and Data Entry Operators came to the households in person and performed their SECC responsibilities?; and Question-6: how the SECC was perceived/ interpreted by the villagers and urban population? To demonstrate the outcome of such an exercise, see Figure 4.2.

Publication of Final List in Public Domain

The SECC final list will be published and put up in public domain as decided by the State/UT governments, including at least in the respective Gram/Urban Panchayat besides web enabling the same in *www.http//:secc.gov.in* for the public to have access. Flow chart for SECC Final List Printing and Publication is given for easy glance (See Figure 4.3).

Conclusion

The introduction of e-modules at all levels and various other processes in conduct of SECC 2011 are welcome initiatives. It can be seen that at all levels, concerns of transparency, verifiability, accountability, public participation and explicability are meticulously envisioned. As an outcome it is expected that at all levels of conduct these key principles are ensured. What remained as more challenging task is to fix accountabilities in the whole processes involved and it has remained passive in the various nuances of administration enforced in governance by stakeholders. A penalty concept over accountability clauses is nowhere insisted in SECC procedures and conduct. This is probably due to giving respect to the federal democratic structure of governance existing in India. Let us await the real field level challenges that may come up as one starts using the SECC data for socioeconomic development. Cross verifications at sub-unit levels to highlight the union and intersections of data specific categories and parameters to establish internal consistencies would throw up errors and disconnects to be resolved via process of grass root level identification and verifications of beneficiary and case specific information. As a follow up action, it would be appropriate to reassess the emergent performance of SECC in achieving transparency, accountability, verifiability and explicability in the e-governance scenario currently on in India via the introduction of TIN (temporary identification number) of SECC/NPR and the Jan-Dhan-Aadhaar-Mobile (JAM) of Digital India.

Figure 4.1

Sample Public Notice

Logo of Govt. of India	Logo of State Govt.

Socio Economic and Caste Census (SECC) 2011

NOTICE

It is brought to the notice of all households that their Socio Economic data enumerated during **Socio Economic and Caste Census-SECC 2011** (except data on Caste) would be put in public domain unless they submit a written request in the format given below that they do not wish to share their data in public. If they do not submit a written request within 15 days to designated **Charge Officer** (BDO/Tahsildar/Sub-Collector/Other) for the SECC 2011 from the publication of the notice that they do not wish to share their Socio Economic data in public at the time of publication of the draft list within the next 15 days to the Charge Officer, it would be presumed that they have no objection to share their data in public at the time of publication of the draft list.

Application Format

To,

The Charge Officer, (SECC, 2011)

Postal address

Sir/Madam,

I.................., S/o, D/o/, W/o......................resident of............Village/town ...., GP................, Tehsil..............,District do not wish to share the household information / data collected during SECC 2011 enumeration at the time of publication of the draft list and thereby in the public domain . I am fully aware about the implications of making such a request.

(Signature or Left Thumb Impression)

Name:

Household Identity No. in SECC 2011..........(not compulsory)

Date:

Place:

Figure 4.2

Results on Questions 1 to 6

Question 1

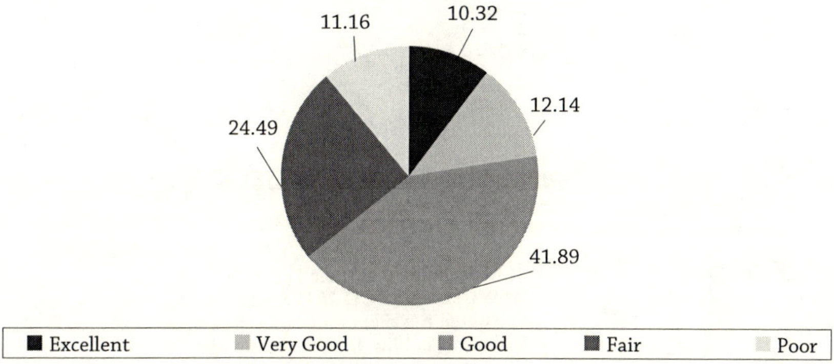

Question 2

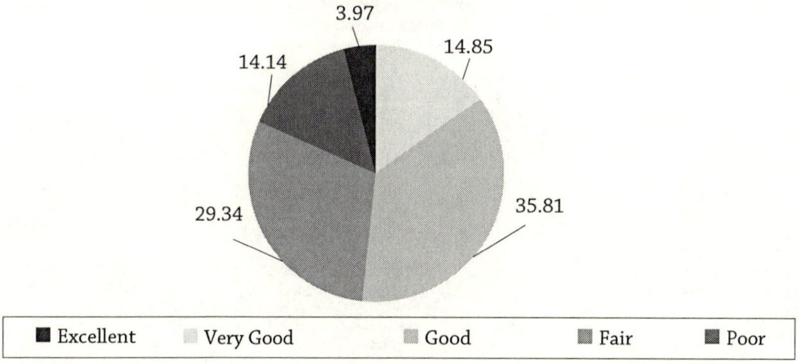

Question 3

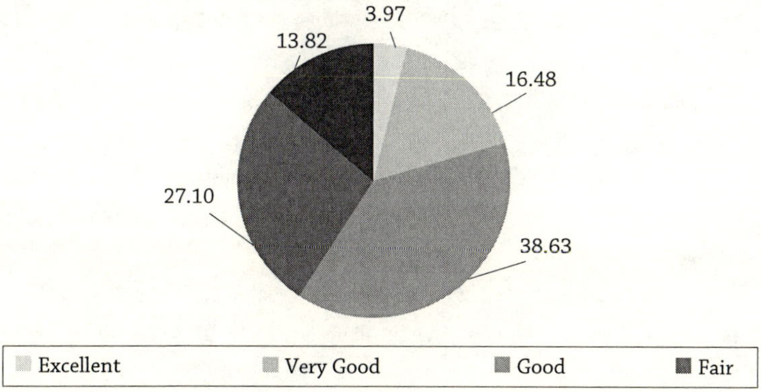

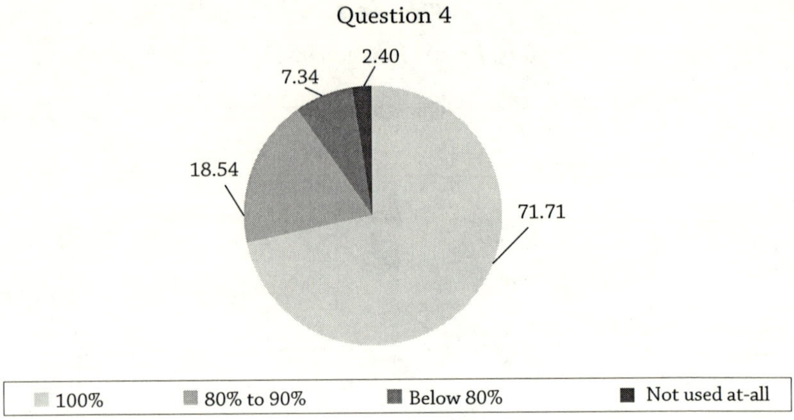
Question 4
2.40
7.34
18.54
71.71
100%
80% to 90%
Below 80%
Not used at-all

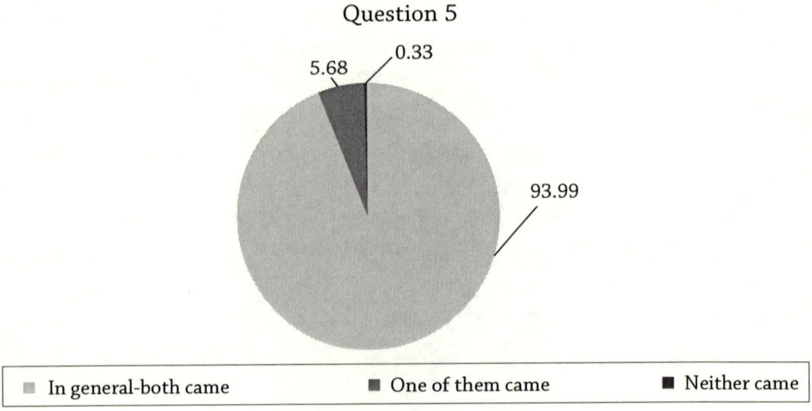
Question 5
0.33
5.68
93.99
In general-both came
One of them came
Neither came

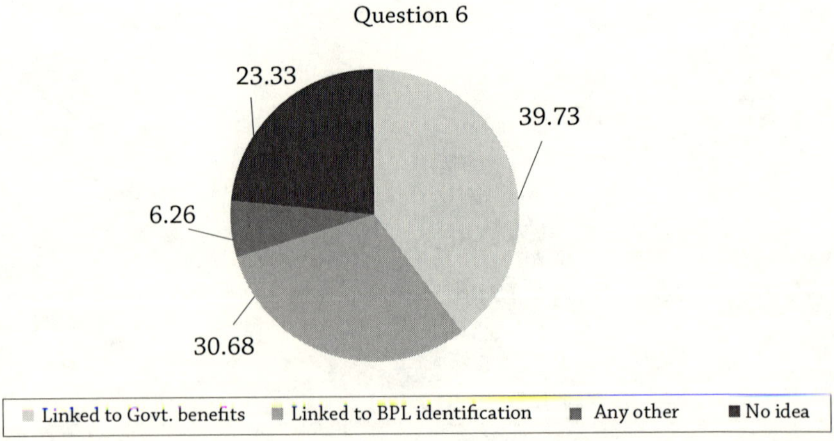
Question 6
23.33
39.73
6.26
30.68
Linked to Govt. benefits
Linked to BPL identification
Any other
No idea

Figure 4.3

Flow Chart for SECC Final Listing Printing/Web Publication

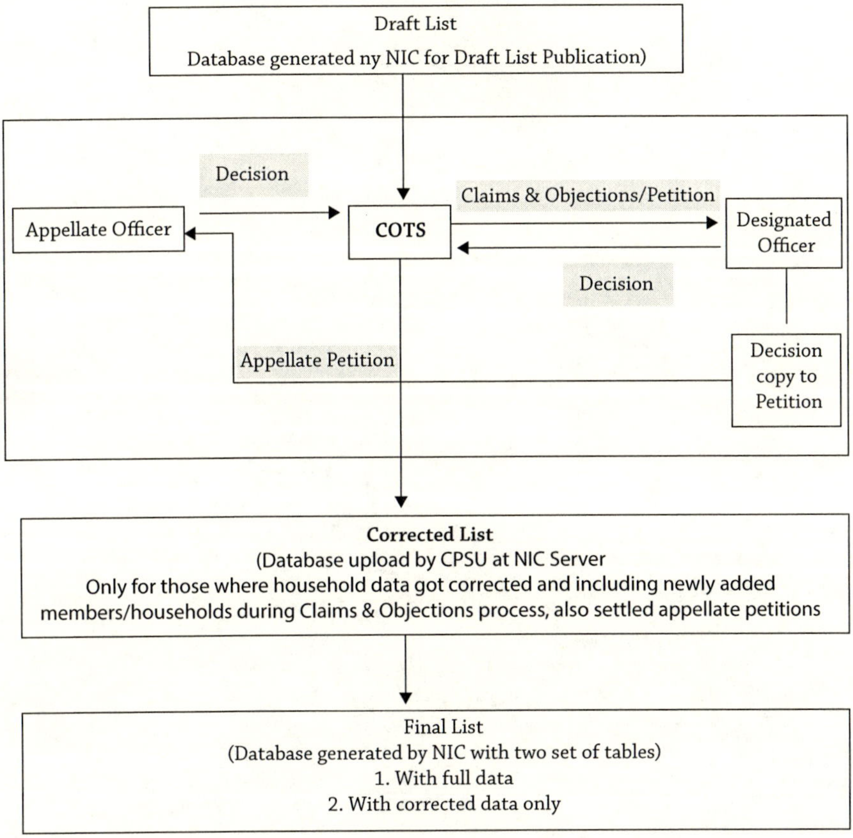

Part 2

Project Management

5

N.K. SAHU

Stakeholders Galore

Kurukhestra was the war between Pandavas and Kauravas—the two powerful royal clans of the epic *Mahabharata*. Thousands of warriors from all parts of the Aryavarta (land mass that includes present-day India that is Bharat and beyond) participated, willingly or unwillingly, directly or indirectly, taking-side or independently. All sections of the society took part either actively or passively. Stake was high for a few, but the remaining fought knowing very well that the end would be apocalyptical. Primarily the fight was for controlling the kingdom—the Hastinapura. Droupadi or Panchali, the wife of Pandavas wanted to take revenge and tie her hairs shucking it with bloods of Kauravas as they insulted her in front of the royal congregations of the Hastinapura. Thus she wished for war. Shakuni, the dear maternal uncle of Kauravas wanted war so that he could see death of all hundred Kauravas as for him they were the cause of the perishing of his parents and family. Though closely in the side of Kauravas', he worked from inside for the defeat and death of Kauravas. Karna the sixth Pandava was in the war as Duryodhana the eldest Kaurava was his dear friend. Similarly, all those who participated in this war has either individual or collective stake. Though difficult, fathoming the intentions of all these warriors, it is not possible to judge the many dimensions of the war. Even the metaphysical and literary value of the epic *Mahabharata* can be understood without identifying and analysing the role played by each and every stakeholder. In present day business management, social engineering, state craft and many other fields, it became imperative to engage in stakeholders' analysis in order to manage the disposable resources and achieve the objectives of the institutions, systems and its members. Thus it became imperative to undertake a stakeholder's analysis of the Socio-economic and Caste Census (SECC) to manage such a complex project and arrive at the intended outcomes. In this chapter we would identify the: (i) main stakeholders, (ii) analyse their stakes (power, influence and interest), and (iii) explain the stakeholders' engagement following structured management principles and tools.

Main Stakeholders

As the stakeholders analysis was first applied to business companies, the purpose here is to borrow the same framework and try to apply it to a project like SECC. The internal stakeholders in a company are employees, managers and owners, and, the external stakeholders are suppliers, society, government, creditors, shareholders and customers. Internal stakeholders are those agencies that are involved in the day-to-day management of the project whose performance determine the outcomes. In case of SECC, they are the: (i) Central/Federal government and its ministries/departments namely Ministry of Rural Development (MoRD), Ministry of Housing and Urban Poverty Alleviation (MoHUPA), and Office of the Registrar General & Census Commissioner, India (ORGI) in Ministry of Home Affairs (MoHA), (ii) state governments and its ministries/departments looking after rural/Panchayati Raj affairs and urban area administrations at state, district and community block/Tehsil, (iii) CPSUs (central public sector units) namely Bharat Electronic Limited (BEL), Indian Telephone Industries Limited (ITIL) and Electronic Corporation of India Limited (ECIL), and (iv) vendors and sub-vendors, either governmental (National Informatics Corporation and National Institute of Smart Government) or non-governmental (private/public limited companies, trusts and societies). On the other hand, the external stakeholders are: (i) general public, (ii) targeted public-beneficiaries of various government interventions, and (iii) government agencies that implement welfare schemes. Though these stakeholders have little or no role in influencing the outcomes of this project, they are directly or indirectly and substantively benefited. The names of main stakeholders are depicted in Figure 5.1.

Figure 5.1

Main Internal and External Stakeholders

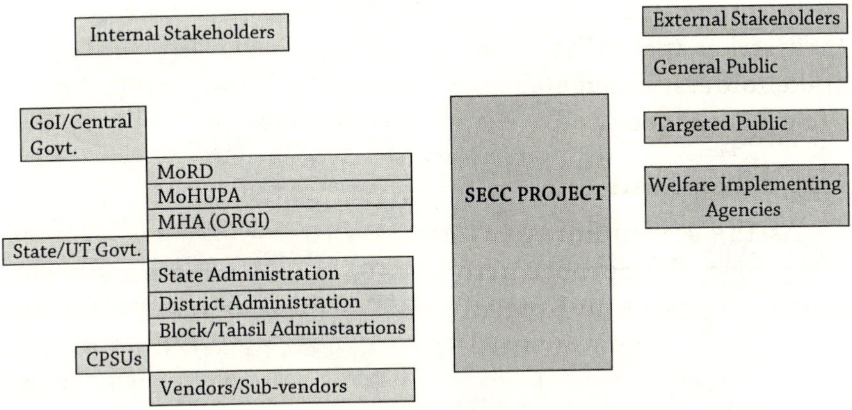

Figure 5.2

Main Stakeholders: Core Circle to the Peripheral Circle

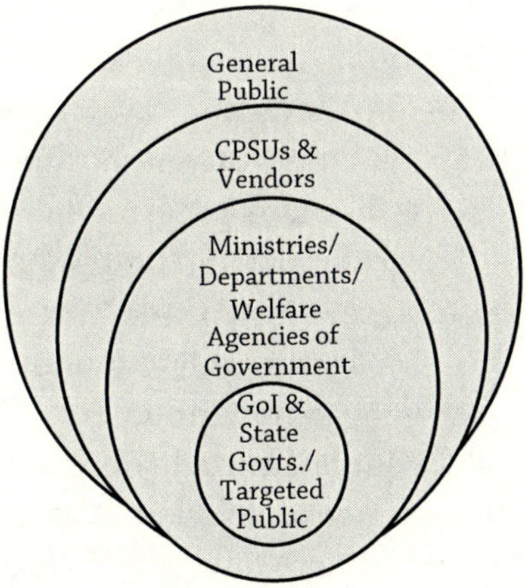

Figure 5.1 and 5.2 are two, out of many, different ways to identify the distinct stakeholders based on their role in the project. While Figure 5.1 classifies the stakeholders as external or internal to the projects key activities, Figure 5.2 classifies them in layers from core circle to the peripheral circle. Internal stakeholders are those agencies, groups or body that work within the project as

managers, financers, monitors, coordinators, resource mobilisers and policymakers. External stakeholders don't manage the project, but are the receivers of benefits or user of facilities or bearer of costs having various stake in the project. Figure 5.2 in its core circle placed the key players that are involved in governance and decision-making process and communicate constantly and intimately with other key players. Briefs on these stakeholders are described below:

a) Government of India (GoI) or the Central government is the core stakeholder as well as the internal stakeholder in the SECC project. India is a welfare state and in order to target the welfare interventions, it strives for achieving efficiency in implementation of its schemes. Since independence, especially with the advent of planned development, many schemes were implemented in various sectors for achieving different objectives. The main sectors are health, education, social welfares of disadvantaged sections of the society like SC (scheduled caste), ST (schedule tribe), women, differently-abled, minorities, backward classes, and weaker/vulnerable sections of the population like, farmers, fishermen, weavers, and unorganised sector workers, etc. Most of the time the welfare schemes are targeted towards economically weaker sections of poor families and/or individuals. As GoI spends millions of rupees for these welfare and development schemes, achieving scheme objectives demands that schemes are well designed and executed. Both at designing and execution stage, importance of identification of targeted beneficiary families and individuals gets utmost importance. Over the years, the GoI is perfecting the beneficiary selection process. Identification of targeted beneficiaries, especially the BPL (below poverty line) or poor families and individuals in census mode in the rural areas was started in 1992 and periodically repeated in 1997 and 2002. Though successful to a large extent, the desire to perfect the art of selecting the deserving population can be seen in these censuses. The GoI designs the census methods, funds the activities, handhold the state governments in survey and enumeration activities. As India is a democratic republic, accountably to parliament, assemblies, PRIs (panchayati raj institutions) and peoples is of utmost value. Keeping all these in view, the SECC project is designed and implemented with GoI and its ministries at its core, assigned direct role in its execution. Though most of the agencies of GoI would reap the benefit of this project as users, the main stakeholders within it are MoRD, MoHUPA, ORGI. While MoRD and MoHUPA implements a number of schemes for rural and urban poor, ORGI in the Ministry of Home Affairs (MHA) was mandated to conduct census of population with their caste category. Caste categories were to be captured during general population census time, that falls in year 2011. Due to some factors, these characteristics of the population could not be captured and thus, this limited but crucial data is planned for collection during SECC. GoI assigned the task of coordination and execution to MoRD as the nodal ministry, the MoHUPA and ORGI played important role in the whole SECC project as core and direct stakeholders.

b) There are 29 states and 7 UTs (union territories) in federal India and all of them were core and direct stakeholders of SECC project. They not only implement all welfare and developmental schemes of the GoI as partners, but also implement a number of schemes of its own specially designed for its target groups. Over time, these states' financial strengths increased and also a large amount was devolved in favour of them by the respective Finance Commissions setup following constitutional mandates. As the state governments are relatively close to people in comparison to GoI, their stakes are very high in so far as the implementation of the scheme is concerned. The state government headed by the council of ministers and the chief minister is accountable to peoples through the state assemblies like the GoI headed by council of ministers and prime minister. In an electoral democracy, accountability to people is sine-quo-non and outcomes of proper execution of schemes result in electoral dividends. Secondly, the SECC project is directly implemented by the state governments. They recruit the field enumerators, supervisors, verifiers, district and state officials and put them to work after training in the census operation with distinctly assigned duties, responsibilities and accountability. The quality of outcomes of the SECC depends on the way this workforce is managed. Within state government, though part of the administration system, the stakeholders are: (i) state administrations, (ii) district administrations and (iii) sub-district (block/tehsil) administrations. These adhere to power, administration and geographical hierarchies, but, play as distinctive stakeholder entities. It is observed that though the structure of power, administrative and geographical hierarchies is mostly similar in majority of states, as separate stakeholders they vary across substantially. Jammu & Kashmir is a state located in north of India bordering Pakistan, Afghanistan and China. It enjoys special status under Article 370 of the Indian constitution. The Vth and VIth schedule areas as mentioned in Indian constitution follows specific administrative setup due to considerable presence of ST population in these areas. The seven UTs are directly under administrative control of GoI. Thus understanding the nature of these stakeholders and designing the implementation architects of SECC is crucial to its success.

c) The targeted public under various welfare and development scheme are core stakeholders to SECC project as selection of them as beneficiary based on this survey could influence their life in perceptible ways. They aspire to get a subsidised house, monthly pension, health facilities, scholarship for pursuing education and getting foodgrains for consumption at subsidised rate etc., to name a few. Getting themselves selected as beneficiaries is probably the greatest stake they have in this project, but, their control over the outcome of such a project is very distance. They are basically external stakeholders. In successive BPL census and this SECC the other stakeholders (GoI, central ministries, state governments, etc.) put in place more rigorous selection processes by advocating transparency and people's participation in a structured framework.

d) Welfare scheme implementing agencies are core stakeholders, but are external to the SECC project. These agencies may be pure government or parastatal of the governments. The pure governments like ministries, departments, districts and sub-district administrations implements a number of schemes, so also parastatal like boards, commissions, cooperative organisation and government corporations. Sometimes the benefits are made available to targeted public in financial, material and mixed forms indirectly (through stages—geographically and institutionally), and sometimes benefits are transferred directly in monetary or value quantum in to the bank/post office account numbers of the beneficiaries. Whichever may be the benefit transfer mechanism and hierarchy, the welfare and development implementing agencies have an important and responsible role to play. Thus they are core stakeholders, but, without power (theoretically) to influence the beneficiary selection process as they are external to the SECC process.

e) Three CPSUs namely BEL, ECIL and ITI were engaged by government to perform major role in the SECC process. They provide all TPCs (tablet personal computers), solar chargers, additional batteries, data entry operators, data administrators and technical support services for the project. Being commercial entities, they received the price for each of these items. Earning profits is the main concern of these CPSUs, but being under the ownership and management control of the government (the core and internal stakeholder of the project) they play a peculiar role as internal stakeholder. Aggregate financial stake was high, but activity wise margin may vary depending upon the complexity and composition of the respective activities. Thus scope of trade-off amongst activities and deliverables do exist and the CPSUs were free to assign separate weight to them, off course with the broad performance parameters setup by the customer (government in this case). These CPSUs engaged hundreds of vendors all over the country and the deliverables are required in agreed quantity and quality at the predetermined palace (at state capitals or at district/block headquarters or at village and town level) and time (during enumeration periods or in the full SECC process periods—from start to end). As the SECC is carried out in lakhs of villages and towns all over the country, the enormous variations encountered by these stakeholders is very crucial to their bottom lines. Sometimes one have to look at their deliverables as public goods and the standard manage principle followed by commercial entities may not be applicable. As internal stakeholders, these CPSUs stand at the periphery of the project.

f) Lastly, the general public is also a stakeholder, though they are external to the SECC project and operate at the remote periphery of the project. General public as taxpayers and responsible citizen expert that the SECC project is well conducted and the outcomes of the project are properly utilised for the betterment of the country. General public plays a crucial role in SECC project during the public scrutiny by participating intimately. The quality of the outcomes depends, besides other, on the people's participation.

Power, Influence and Interest of Stakeholder

After identification of the stakeholders following set principles (external-internal and core-periphery) in the previous section, it is proposed here to analyse the power, influence and interest of the stakeholders in this section. The standard tool to place the stakeholders in a power/interest grid is followed (*https://en.wikipedia.org/wiki/Stakeholder_analysis*) and it is depicted in Figure 5.3.

Figure 5.3

Power Interest Grid: Placement of Stakeholders

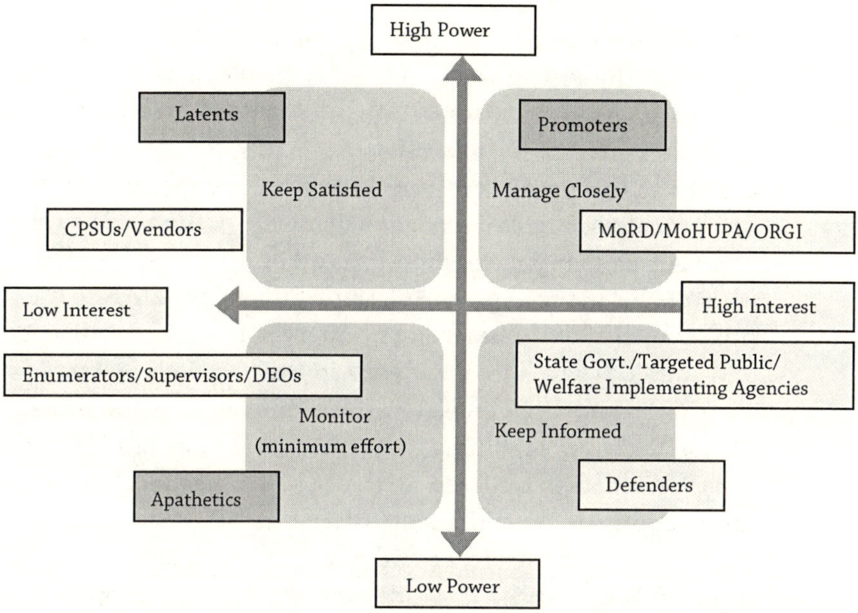

a) *Top-right box*: The stakeholders placed in this box are the promoters like the GoI (MoRD, MoHUPA and ORGI). They have not only considerable powers over the SECC project, but also high interest. From the project conceptualisation stage to bring it to a logical conclusion, at all stages these agencies worked intensively with a clear objective and purpose. The census methods were finalised with the constitution of Expert Committees in MoRD and MoHUPA for rural and urban areas respectively. The concepts used and the methods of survey were finalised after rounds of discussions with the main stakeholders like state governments and the user ministries/departments. GoI funded it fully and managed the finance following the government rules and procedures. How to use the outcomes of the census and update the data is also decided by the promoters, of course after due consultation with the data user stakeholders. Close management of the project was assigned to the nodal ministry—MoRD, and the other promoters play their due role from start to end.

b) Bottom-right box: The stakeholders in this position have lots of interest in the project, but very low power to influence it. We may put the state governments, union territory administrations and the targeted public in this box. They have high interest on not only the project itself, but the outcomes of the project. As outcome depends on the process, structure and the management of the project, the state governments have high interest in these aspects. They control the nature and quantity of human resource for the project—the field enumerators, supervisors, verifiers, district and state officials, of course with in the broad parameters as per the project guidelines. Once the outcomes of the project are handed-over to states, they are supposed to use it for all types of schemes targeted at the beneficiaries located in its jurisdictions as per the constitutional framework. Credibility of the government are at stake, as proper selection of the targeted beneficiaries is conditional upon the quality of the SECC project outcomes. Though the interest of the state government is high, its power is relatively low with regard to either the management of the project or to influence the outcomes. Instructions or guidelines for all types of participants and processes were issued by GoI and funds were provided to the states. For such stakeholder, the project demands achieving symmetry in full information flow, meaning, to keep them fully informed. Throughout the SECC project period, the main promoters' organised PRC (performance review committees) meetings every quarter with the secretary MoRD in chair. Very frequent VCs (video conferences) were organised to mainly short out day-to-day problems. Sometimes it is with as large number of states and some other times it is with only a few states. The public website is designed for SECC to keep informed all the stakeholders including the states. The other group of stakeholders that have high degree of interest on the outcomes of the project, but have any power to influence its management and/or outcomes are the targeted population and the welfare implementing agencies. Being not in the management of the project, they are powerless though fully interested in the outcomes as the possibilities of influencing the life and well-being of targeted public depends to large extent on the outcomes of the project. The strategy for the project managers is to keep this group informed at each stage on all aspects. The 'claims and objection process' before finishing the SECC project is one such method that aims at keeping these stakeholders informed, besides it aims at bringing transparency and credibility to the data. These are the stakeholders, who would act as defenders of the project and the outcomes because of the very role they play in the whole process.

c) Top-left box: The stakeholders in this position have low interest in the project, but very high power to influence its management and outcomes. CPSUs and their vendors work on commercial principles. SECC project having social and political objectives of identification of target beneficiaries and facilitating in directing the government schemes and interventions to these sub-set of population have any or little interest to these stakeholders. Even though the CPSUs are owned and managed to a large extent by government, its' core concern is not social well-being of the disadvantaged and vulnerable sections of the society. Similarly hundreds of vendors and sub-vendors engaged by these CPSUs being commercial entities, subscribe hardly to the project objectives. All of these were working for a price and commercial viability of the deliverables is the key concern. The purpose for which the promoters work is not the same for these stakeholders. One thing is to be kept in mind that there are two parts to this SECC project, one is its management and the second is the outcome of the project. While these stakeholders may not have much interest in the outcomes of SECC, its' interest in the management of the affairs is of considerable importance. As because, the efficient management of this project means scope for earning profit for the CPSUs and its vendors. It is another thing that these arrangements does not hold good in this project as the project length goes on stretching beyond the targeted time frame by many times. The six month project time was extended beyond four years cutting in to the profit margin of these stakeholders. Secondly, the Top-Left box also depicted that these stakeholders are quite powerful. The CPSUs are powerful as they are big public sector units owned by GoI and even few of the vendors are owned by the state governments. They are few in number and thus capable of wilding considerable power due to monopoly positions, despite being owned by governments. These CPSUs also have powerful influence on the management of the project due to their proximity to the promoters. They were working closely with ORGI for National Population Register (NPR) project before being hired for SECC. Because of all these circumstances, these stakeholders are described as 'latents' and are kept satisfied, though not in financial terms but in congenial working environment terms.

d) *Bottom-left box*: In management parlance, these stakeholders are called 'apathetic' as they are low in both power and interest in so far as the SECC project is concern. They are the thousands of human power (enumerators, supervisors, data entry operators, technicians etc.) working for the project, and to a great extent the general public. Once in five years the general public get power to elect a government under the constitutional framework and keep themselves busy in rest of the times in day-to-day life, remotely showing any interest in improving the life of the left-out folk. For the management point of view of this SECC project, these stakeholders are to be closely monitored with the objective to extract the services required from them. In so far as the general public is concerned, they need not be monitored or controlled as they are remotely located stakeholders.

Nature of the Stakes

In this section, our purpose is to analyse the nature of all the identified stakes as well as the stakeholders. As the stakes and the stakeholders are two sides of the same coin, we would explain the nature of both together side-by-side. This will be done with the help of the analytical framework developed to analyse the working of the Members of Parliament, Local Area Development Scheme being implemented in India as a centrally sponsored scheme (Sahu 2006). In this framework (both tabular and diagrammatical), "the stakes are measured in three types of relative scales. They are, (i) direct to indirect scale, (ii) private/personal to general/public scale, and, (iii) high to low intensity scale.....If the intensities are high, the location of the stake will be in one extreme, but if the intensities are low, the location would be would be in the opposite extreme" (Sahu 2006: 52). This framework is little modified to suite the SECC project. There are 13 stakeholders and a minimum of about 17 number of stakes. In Table 5.1, the major stakeholders and the name of the stake is depicted. In order to understand the nature of these stakes, these are diagrammatically depicted in subsequent sections.

Stakes

There are many individuals and entities (business house, government organisation, non-government organisation—for profit and not-for profit, civil society organisations, etc.) affected by an action. The action may be in the sphere of social, political and economic environment—exclusive of each other or in combinations. The individuals or entities are affected positively or negatively. Even sometimes the effects are neutral. The main 'stakes' are explained here to understand the nature of each one. It is difficult to serialise the stakes following any established principle, and thus, these are explained randomly. As each stake is a stand-alone case, it is advised to understand it in relation to its' stakeholder alone, not in relation to any other stake.

Table 5.1

Stakeholders and Stakes in the SECC Project

	Stakeholders	*Stakes*			
1	General Public	Govt. scheme benefits			
2	Targeted Public	Govt. scheme benefits			
3	Central Govt.	Public welfare			
4	State Govt.	Public welfare			
5	Welfare Implementing Agencies	Easiness to execute	Manipulation difficulty	Pecuniary/ Non-pecuniary benefits	
6	MoRD	SECC quality	Beneficiary selection	Financial prudency	SECC completion
7	MoHUPA	SECC completion	SECC quality	Beneficiary selection	
8	MHA (ORGI)	SECC completion	SECC quality		
9	CPSUs	Pleasing govt.	Profit	Quality deliverables	Timelines
10	Vendors/ Sub-vendors	Profit	Pleasing client CPSUs	Quality deliverables	
11	State Administration	Nodal activities	Managing human resources	Financial management	SECC quality
12	District Administration	Nodal activities	Managing human resources	Financial management	SECC quality
13	Block/*Tehsil* Administrations	Nodal activities	Managing human resources	Financial management	SECC quality

a) *Public welfare*: The SECC was conducted by the government in order to target its interventions for the welfare of general public. In a democratic nation like India, the constitution in Part IV spelled out the Directive Principles of State Policy that mentions about promotion of welfare of the people and specifically in article 41 that ensures right to work, to education and to public assistance in certain cases. Thus since independence, the governments, both at union level and state/federal level follow policy of social welfare and SECC and BPL Census is devised to act as an effective tool to achieve this goal. This tool is useful at the time of planning and execution of welfare schemes. Census provides vast data on various features of the population and society, which is utilised to arrive at policy decisions and implementation architects.

b) *Govt. scheme benefits*: Union, state and local governments implements a number of schemes, programmes and interventions offering various type of benefits to the targeted public. Those benefits may be old-age pension, food security, subsidised gas cylinder, scholarship or tuition fee weaver for students, house and financial or other types of benefits to disadvantaged, to name a few. Given the poor, powerless and ignorant nature of sizeable population in India, it is expected that getting government scheme benefit is priority for them. If SECC is done following scientific principles, in transparent and participatory ways, the deserving and aspiring public would feel confident of their selection. Thus this stake is of utmost importance to such aspiring public in particular and common public in general.

c) *Easiness to execute*: When an agency is assigned to work as conduce between a scheme beneficiary and the benefactor, the requirement is an effective implementation mechanism. Simple the better. Less work the better. Thus the easiness in implementation of a scheme is aspired by the executing agency. If the most difficult task of identification of beneficiaries is taken care of SECC, the governance of these schemes become relatively easy.

d) *Manipulation difficulty*: The implementing agencies are composed of humans with its all fallible natures. Some may be interested in totally non-manipulatable beneficiary process, where others may be interested in manipulating the process. The balance is not easy to predict. However, 'manipulation difficulties' is a major stake for welfare implementing agencies.

e) *Pecuniary/Non-pecuniary benefits*: When the implementation mechanism and the rules and regulations are unfathomable it becomes susceptible to manipulation. The officials that are assigned the task to select the beneficiaries and distribute the scheme benefits looks forward to pecuniary and non-pecuniary benefits. The outcome of such a system is unethical, though it is prevalent in present day India.

f) *SECC quality*: MoRD, MoHUPA, MHA (ORGI) and state/district/local administrations have taken the responsibilities to conduct the SECC and achieve the maximum possible quality of the outcomes. The question is not whether these agencies are convinced about the high quality required out of this project, but it is the management principles that requires them to aspire for quality outcomes. To achieve quality, a number of steps were decided prior to the start of the project and during the project. Though quality of SECC is an abstract phenomenon, strictly following the laid-down procedures to the fullest is what these agencies tried to achieve in order to claim the stake—SECC quality.

g) *Quality deliverables*: Here the deliverables are something that are expected from the CPSUs and the vendors/sub-vendors of the project. These are all documented and agreed to, before the work is assigned to them by the clients (MoRD, MoHUPA, ORGI, States etc.) and are quantifiable. The tolerable limits of quality for each deliverable is achieved. Otherwise, there is a prescribed course correction protocol and even in some cases financial penalties. It is therefore the CPSUs and vendors that tried to achieve the required quality, and thus the main stake for them.

h) *Beneficiary selection*: A poor old person above 60 years has to get a fixed amount as pension each month. In some states, the full amount is paid by the union government and in others, an additional amount is paid by the state government. Even in few states, this benefit is targeted to population relaxing the age and BPL criteria. In all these cases, the challenge is to select the beneficiary in a normative and transparent method. There are hundreds of such schemes where the target beneficiaries are different sets of people or households. If the targeted beneficiaries are not properly selected, the scheme resources flows to undeserving and its implications may go beyond finance. Social and political implications may be detrimental to the peace and harmony of the society. Therefore the governments, both union and state, works under the constitution to select the targeted beneficiaries as scientifically as possible.

i) *Financial prudency*: This stake is SECC project specific. A fixed amount of financial resource was set aside to conduct the SECC, and, the money is to be spend following sound finance management principles. As this is a public project, the GFR (general financial rules) as adopted by the GoI would be applicable. The project managers aims at achieving prudency in financial management.

j) *SECC completion*: The latest SECC like census conducted was in the year 2002. That was conducted by state governments and union territory administrations with technical and financial support of MoRD. Completing the BPL Census 2002, limited to only rural areas, took an arduous and long path occurred due to court cases. The implications of such delay is enormous and the data base created through this is been used even today to identify beneficiaries under rural development schemes. The government learned a lot from this exercise and adopted ways to avoid delay. For example, no court case came on its way and thus further delay was avoided. This could be done by constant interaction with the potential litigants and designing the processes carefully to avoid legal complications. The communication mechanisms were designed to establish close and intimate flow of information amongst all the major stakeholders. The vastness of the coverage of the project and the complexity of the sociopolitical mosaic was kept in mind at each stage to avoid delay. All stakeholders, including MoRD, MoHUPA, ORGI, worked relentlessly to complete the project.

k) *SECC quality*: As selection of beneficiaries under various schemes of government is the prime objective of SECC, high quality outcomes ensures correct selection of beneficiary. The deserving individuals and households under a scheme are to be identified based on the data and parameters collected in SECC. Selection of undeserving is to be avoided as far as possible. The arriving at high quality outcomes of SECC is of utmost importance. Any dilution of this stake means inviting infectiousness. General public including targeted public has stake in quality of SECC, but the stake is paramount for MoRD, MoHUPA and state governments.

l) *Pleasing government*: Union government and state governments are the owner of the CPSUs and the management of these depends to a large extent on government. The chief executives of these CPSUs are appointed by government. Thus working as per the directions of government is what these CPSUs do. Even though these are supposed to be managed on profit maximisation principles, in the name of social responsibility, they have to run as directed by its owners, the government.

m) *Profit*: The CPSUs engaged a number of vendors to outsource various technical services. These vendors are either public sector units or in private sector. They provided about six lakh technically qualified DEOs (data entry operators), DBAs (database administrators) and engineers. All for a price and thus to earn profit. Three CPSUs and hundreds of vendors and sub-vendors managed their disposable resources—man and material judiciously with an objective to maximise profit.

n) *Timelines*: There are a number of processes to be followed from the start to end of the SECC project. Each stakeholder has specific role in each stage. The deliverables are to be delivered as per the stages and its' timelines. Many of the stages/processes are sequential, but few of them are concurrent also. Thus maintaining time schedule is of great importance to these stakeholders. Financial returns depends on adhering to time schedule. While completing the whole

SECC may be the main stake for government, completing each stage/process are of great importance to CPSUs and the vendors.

o) Pleasing client CPSUs: Like the objective of CPSUs is to please the government, the objective of the vendors and sub-vendors is to please its clients—the CPSUs. The business environment is to be so built that the vendors get further contacts in future. For this they have not only to deliver the deliverables in required quantity, but also in agreed quality timely. Besides, establishing a conducive and warm relationship with the client CPSUs is also the lookout of each vendors.

p) *Nodal activities*: Like MoRD is the nodal ministry to overlook the whole SECC process in the country, the state governments, district and block/tehsil administrations are nodal agencies at state, district and block/tehsil level respectively. Mobilising human and material resources and putting them to use is the responsibility of these nodal agencies. Though these nodal agencies have to follow the instructions of the SECC processes, they have enormous power and responsibilities to supervise all the participants (enumerators, supervisors, verifiers, DEOs and database administrators, etc.) and take due actions to meet the project objectives. Whenever there is shortage of secondary and higher-secondary school teachers to act as enumerators, these nodal agencies engaged suitable manpower as substitute to recommended manpower. As these nodal agencies are located near to the action locations, they are most suitable to take appropriate actions.

q) *Managing human resources*: SECC being a massive exercise, demanded enormous resource, mostly human resource. Target is to capture socioeconomic data of each and every household of the country. Each habitation, both in rural and urban locations have to be visited by trained persons to collect data. The estimated requirement for such manpower runs to more than a million (enumerators, supervisors and data-entry-operators) that are to be recruited, trained and deployed by state governments and CPSUs. Similarly thousands of technical persons, coordinators and managers are recruited to provide technical and managerial services at block/*tehsil*, district, state and national level. Managing this huge human resource is the challenge and extracting the desired services from them is the main stake for all the stakeholders internal to the project.

r) *Financial management*: For procuring tablet PCs, making payment to millions of manpower engaged in the project and purchasing stationary etc., the union government provide thousands of crore rupee. This financial resource flowed from the exchequers of union government to the persons engaged in field and the service providers. Thus managing the finance prudently and following accounting principles is the main stake of the finance managers—MoRD, state governments, district and block/*tehsil* administrations etc.

Stakeholders and Stake: Analytical Explanations

In the above sections we have listed the stakeholders and the stakes and explain them mostly as standalone entities. But now we proposed to analyse them in combination by putting the stakes of each stakeholders from the point of view on: (i) importance of stake (big and small), (ii) direct or indirect stakes, and (iii) public or private stakes. For each stakeholder, we depicted these dimensions in figurative or graphical presentations (Figure 5.4 to 5.15). This analytical tool was self constructed by the present author in Sahu 2006. The abscissa or the horizontal axes (typically x-axis) of our two-dimensional graph have a scale public to private and the ordinate or the vertical (y-axis) have a scale indirect to direct. The size of the circle depicts the importance of the stake—bigger the circle the bigger the stake. It may be noted that scales of the figures are ordinal and relative. The scale is between zero and seven. There is no logic why the scale is such, in relative terms the spread is specified like that. These figures were prepared based on studying the behaviour of all the stakeholders, and, may have subjectivity. However this tool is useful through sufficient light to understand the interactions between the stakeholders and the respective stakes.

Figure 5.4

General Public's Stake in SECC Project

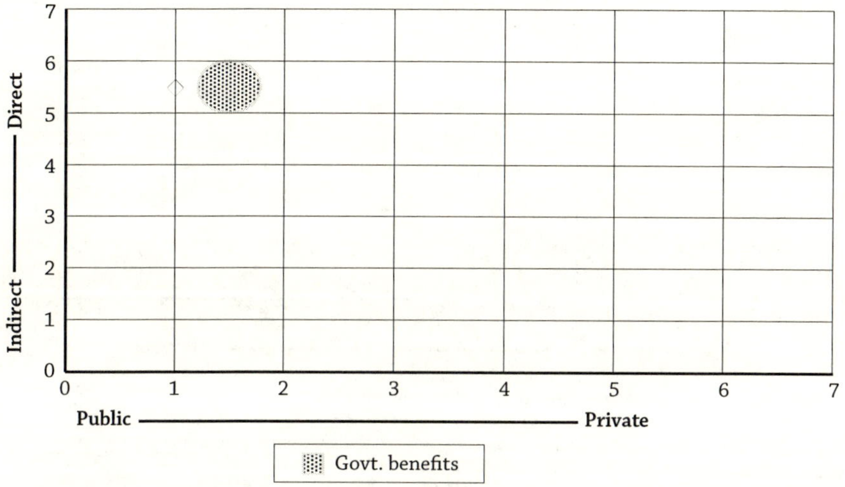

Figure 5.4 shows that the general public have only one identifiable stake, viz. 'government benefits'. This stake is more in public in nature and can be received directly by the general public. General public have multiple of stakes in their life and they do many things to survive. Under such circumstances, 'government benefits' is relatively small stake.

Targeted public is a sub-set of general public who are targeted by governments (union, state and local) under schemes and interventions. Figure 5.5 depicts the nature of the stake, viz., 'government scheme benefits'. Once the stakeholder is selected, it is very much private to the targeted public and comes directly to them. Secondly the stake is very big (circle is largest) as these targeted public are resource poor and needs help of the government to survive as humans and progress.

Figure 5.5

Targeted Public's Stake in SECC Project

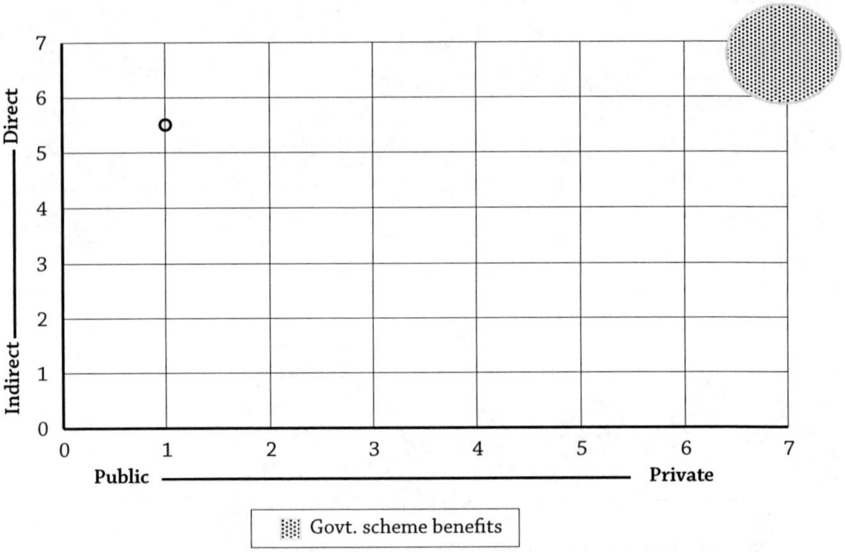

Figure 5.6

Central Government's Stake in SECC Project

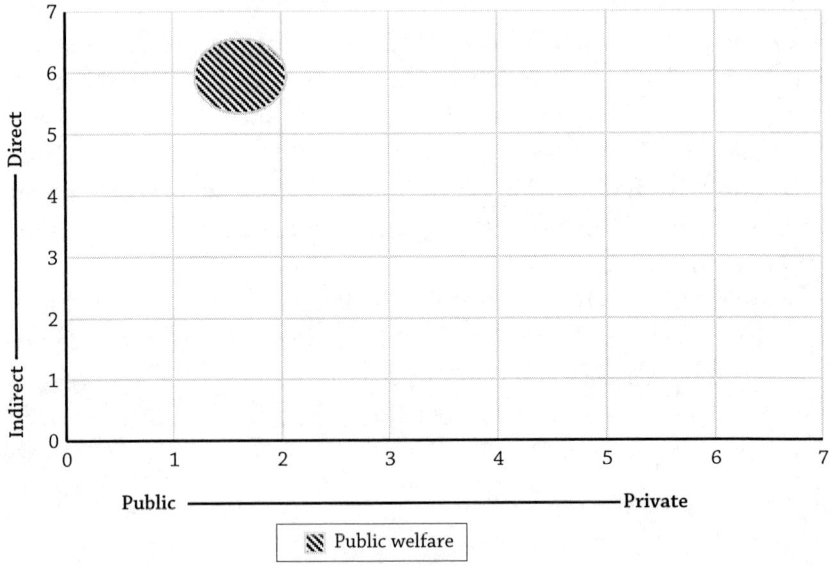

Figure 5.7

State Government's Stake in SECC Project

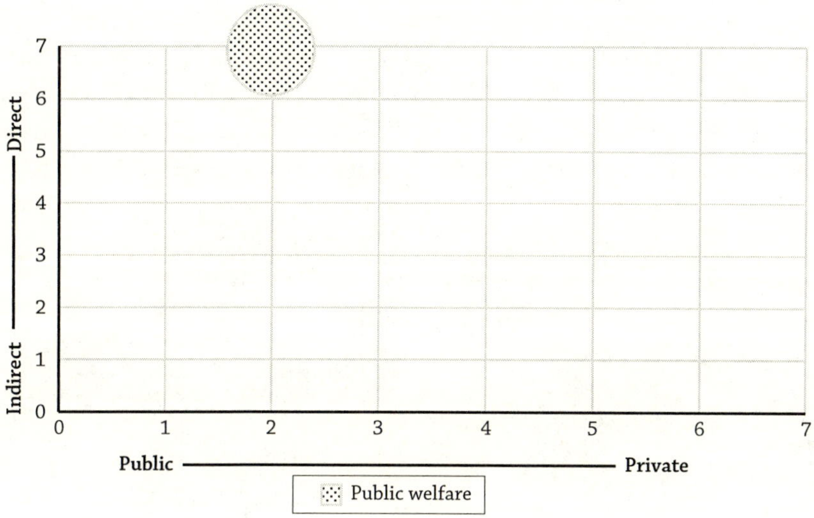

Figure 5.8

Welfare Implementing Agencies' Stake in SECC Project

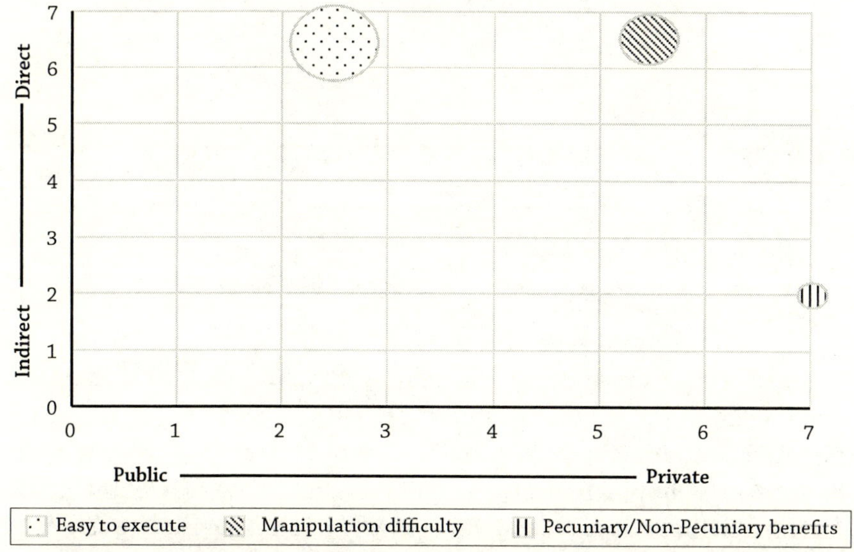

The Central government's stake as shown in Figure 5.6 is mainly the welfare of the public. This stake is directly accrued to public and is mostly public in nature. The stake depicted as a circle is medium in size.

Figure 5.7 depicts the stake of the State government and is very little different from the Central government. This is because both are partners in most of the scheme implementation and funding. However, as the state governments mostly execute the schemes the scheme benefits in terms of public welfare comes very directly.

Figure 5.9

MoRD's Stake in SECC Project

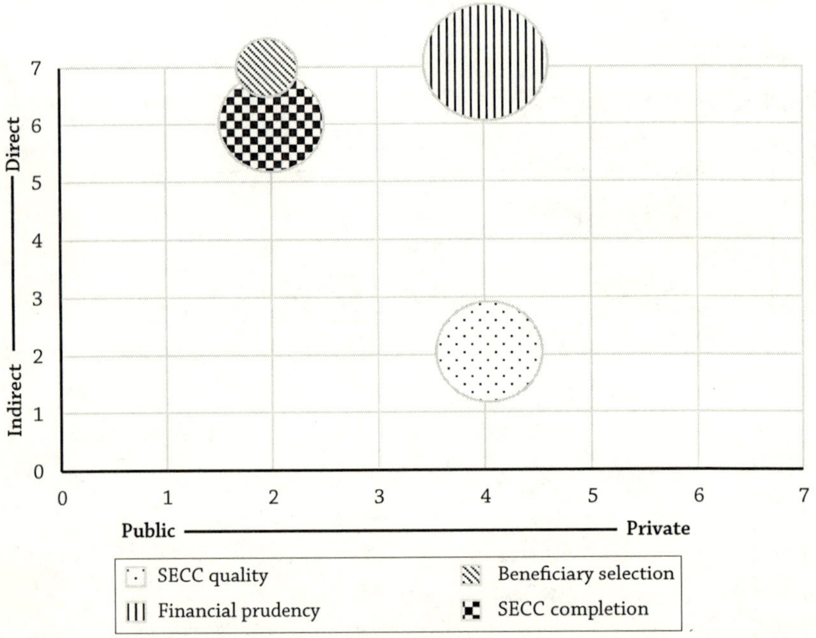

The main stakes of the welfare implementing agencies are 'easiness in governance' of the schemes, 'manipulation difficulties' and 'pecuniary and non-pecuniary benefits'. These are shown in the Figure 5.8. The 'easiness in governance' of the schemes is very much direct to the implementing agencies as this is the last point in the delivery chain and such stake is relatively public in nature. Such stake is also very large for the stakeholders as easy to govern means less work and tension—a welcoming aspect of government employees. Though the scope of manipulation of the size and directions of benefits do exist for the implementing agencies, it is mostly private in nature to them, though directly under their control. Where ever scope for manipulation exists, though small, the implementing agencies (or some of its members) could derive some pecuniary and/or non-pecuniary benefits. Such stakes are totally private to the stakeholders and can be derived very indirectly by indulging in lot of manipulations.

Figure 5.10

MoHUPA's Stake in SECC Project

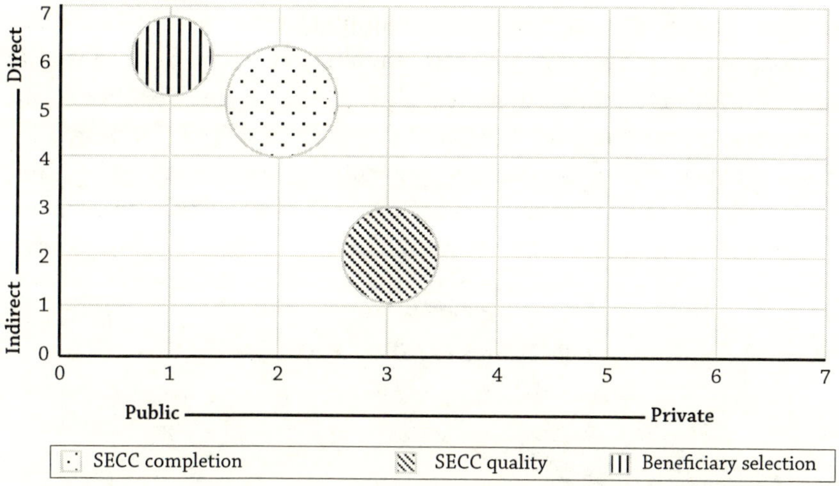

In the Figure 5.9, four stakes of the MoRD are depicted. The stakes are: 'SECC quality', 'beneficiary selection', 'financial prudency' and 'SECC completion'. As nodal ministry of GoI for the SECC project, all the funds are placed at the disposal of ministry of rural development and managing the funds was the prime stake and thus achieving financial prudency was the greatest stake (biggest circle) that directly accrued to MoRD. Completion of SECC and achieving quality are also important (circles of almost equal size), but while the first is a direct stake of public nature, the second could be indirectly achieved and depend more on private initiatives of MoRD. Lastly the fourth stake, viz. beneficiary selection being the primary objective of SECC, is directly concerned to MoRD and is of mostly public in nature. In Figure 5.10, the stakes of MoHUPA is depicted and the explanations are similar to as that of MoRD.

In Figure 5.11, the stakes of ORGI in the MoHA is shown. There are two main stakes, viz. SECC quality and SECC completion. Both are relatively large in importance and can be achieved directly by the stakeholder. As caste data could not be collected during the decennial census, it was tagged with SECC and thus completing the project is the most important stake for ORGI.

The CPSUs have four main stakes, viz. 'profit', 'quality deliverable', 'timeliness' and 'pleasing government' are shown in Figure 5.12. Being owned by government, the management takes all measures to please the government and also earn as much profit as possible (having bigger circles respectively). Quality deliverables and timeliness are smaller, but crucial stakes shown in relatively smaller circles. Both profit and pleasing government are direct stakes and mostly private in nature, but, timelines and quality deliverables are mostly indirectly accrued and of public in nature.

Figure 5.11

ORGI's Stake in SECC Project

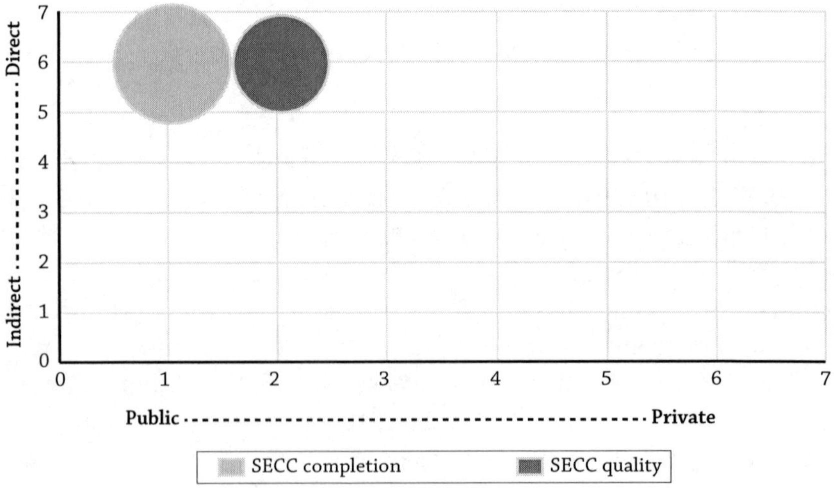

Figure 5.12

CPSU's Stake in SECC Project

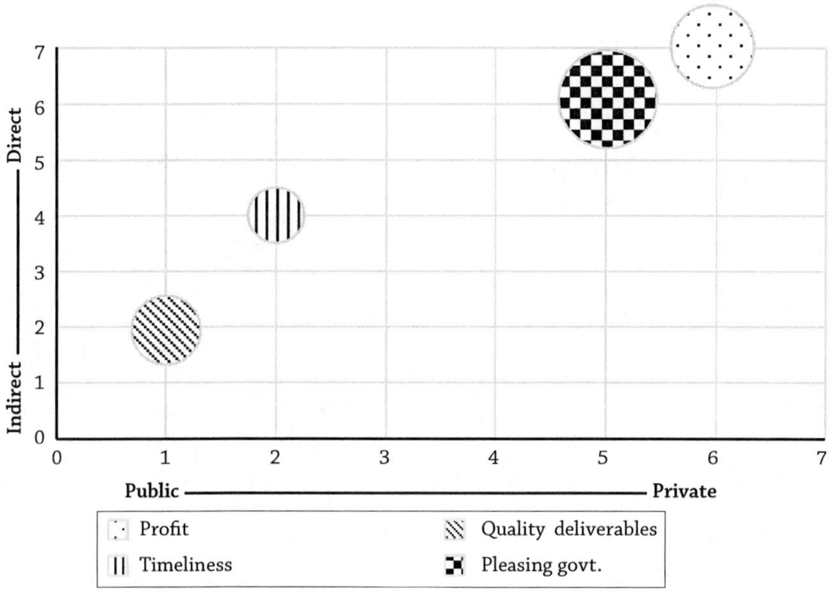

Figure 5.13

Vendors and Sub-vendors' Stake in SECC Project

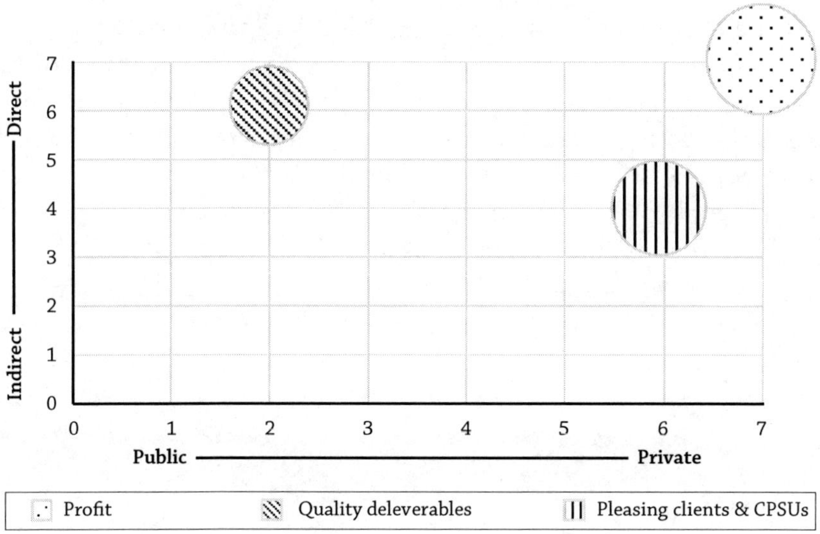

Figure 5.14

District Administration's Stake in SECC Project

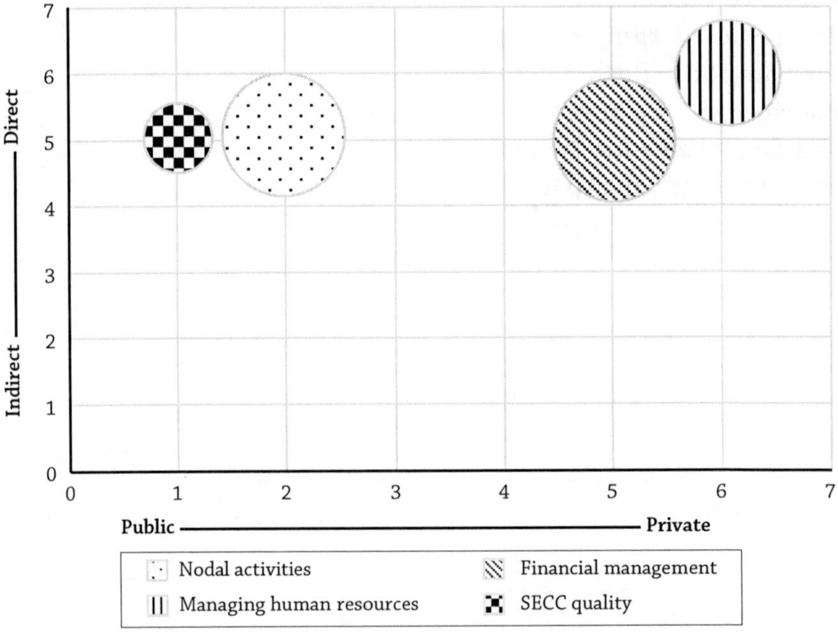

The CPSUs have engaged hundreds of vendors and sub-vendors and the stakes shown in Figure 5.13 indicate that earning profit is the most important stake that is directly under the control of the stakeholder and this stake is private in nature. They are bound by the obligations to deliver quality outcomes over which they have direct control but towards public domain. As these vendors and sub-vendors look for long-term clientele, they try to keep the CPSUs in good humour and for that take steps to please the clients.

The remaining main stakeholders are the State Administration (Figure 5.15), District Administration (Figure 5.14) and Block/

Tehsil Administration (Figure 5.15). They have not only same types of stakes, but also the importance, private-public and direct-indirect nature of the stakes differ a little. They are nodal agency to organise the SECC in their respective administrative and geographical jurisdictions. Both financial management and human resource management are big stakes for all three stakeholders, but are private and direct stake for them. On the other hand, SECC quality is direct stake towards public domain.

Figure 5.15

State Administration's Stake in SECC Project

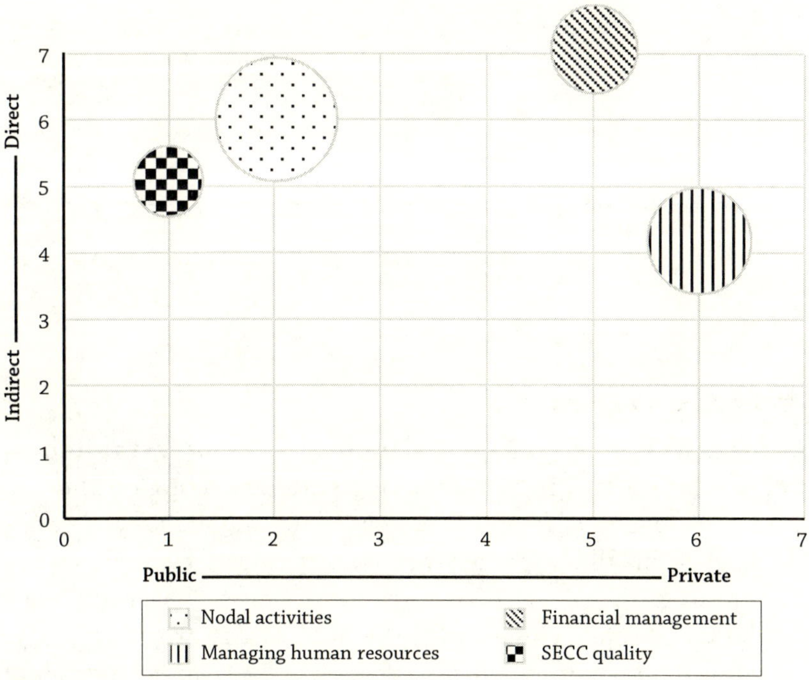

Figure 5.16

Block/Tehsil's Administration's Stake in SECC Project

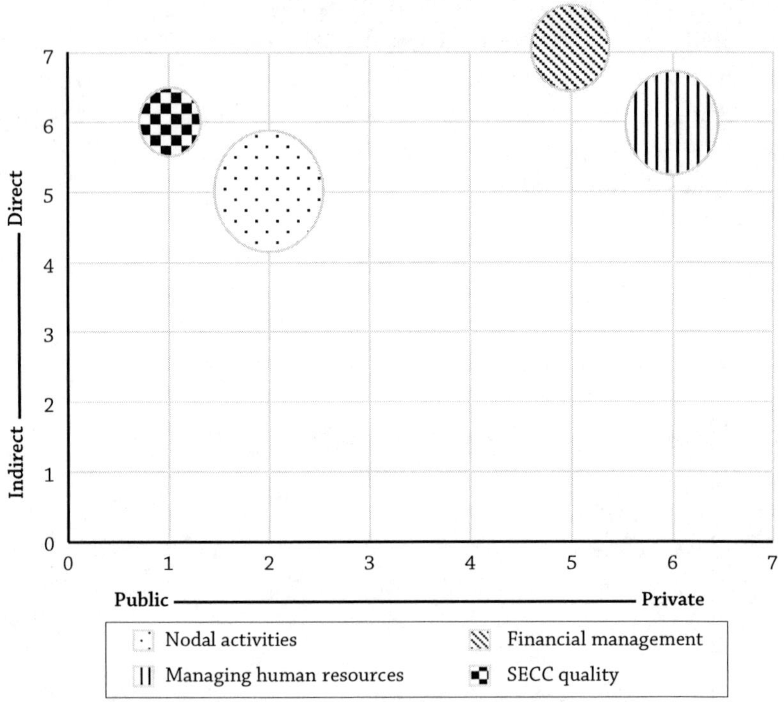

Stakeholder Engagement

Stakeholder engagement is a process of establishing a communication and engagement structure depending on the nature of all the stakeholders and the stakes. The individual stakeholders in a project may be a party who participated either voluntarily or involuntarily. The perceived gains and sacrifices of each stakeholder may be its own judgement, but, the duty of the project manager is to establish an environment of certainty of roles and responsibilities for each of these stakeholders. The stakeholders may be internal or external to the project, or at the core or periphery of the project, but they have the power to influence the outcomes of it. The influence may be small and inconsequential, but ignoring such stakeholders' role may be detrimental to the project. This is more so when the project is implemented in a sociopolitical sphere and the human elements are dominant determinants in the process. In this section, our objective is to explain the communication and engagement structure that worked in this project.

We have explained the nature of internal and core stakeholders in the SECC project in previous sections. They are the union government at the centre, state governments having jurisdictions over its geographical boundaries, MoRD—a part of union government but performing the nodal agency's role to the whole project, other union ministries (MoHUPA, MHA) and CPSUs. The very nature of these organisations, is bureaucratic and rule based. The command and control system is pyramidal, minister and secretary at the top and other staffs below them following a hierarchical structure. The interaction environment is based on command and control based on rules and regulations. As these agencies/organisations work within democratic federalism as enshrined in the constitutional framework, distinctions between the functioning of union government and state governments have to be recognised to implement a project that encompasses jurisdictions of both union and state governments. Secondly, the specific nature of the management structure of CPSUs and its capabilities are duly recognised at every stage of the SECC process. The features of communication and engagement system established for these stakeholders are as follows.

The main stakeholders at union government level are MoRD, MoHUPA and ORGI in MHA. The ownership of the outcomes of SECC and the household data, is compartmentalised. The quality of data in rural area is the primary responsibility of MoRD, urban data is that of MoHUPA and caste data is of ORGI (MHA). But as nodal agency MoRD is overall in charge of the full data set. Before start of the project and during the project implementation time, close face-to-face interaction was the mode of engagement. In all the PRC meetings and VCs, all three participate and becomes party to each decision. The VCs are so structured that MoRD, MoHUPA and ORGI come together in a common location and establish communications with other stakeholders at the other end. Though the instruction manuals and directions are issued separately by them, synergy is established between these seamlessly. This was possible as the nodal officers for these ministries (Economic Advisers in case of MoRD and MoHUPA and Additional Director General in ORGI) communicate to each other continuously and regularly during the project period. Keeping the records of communication is the official requirement in these ministries, and, modern-day technology made it possible effortlessly. Most of the communications were through e-mails and copy of the instructions are posted in the websites.

The communication and engagement structure established between the union government agencies (MoRD, MoHUPA and ORGI) on the one hand and state governments on the other is the most complex one. The political parties ruling the states having different political ideology followed strategies that may not be in conformity to that of the union government. As the ultimate objective of the SECC project is to help and assist the deprived and poor population of the society, the state governments agreed to the idea of conducting the census to identify the targeted individuals and households. Secondly, the union government came forward to fund the project fully. The methods of identifying the beneficiaries were discussed between these stakeholders and the field survey methods were arrived at based on discussions and scientific pilot surveys. During the project implementation time, communications were established through PRC meetings, VCs, special state specific meetings and visit by officials from both sides to each other's locations. Besides hard paper based official communications, e-mails was the main tool of communication. When these stakeholders specific needs were taken care of, the participation of these states become vigorous and sincere. Only a few examples are cited here, but ample evidence is given in subsequent chapters. The data capturing and transfer software were modified to accommodate the state specific requirements (Bengali language is parts of Assam) and hierarchy of data aggregation protocol changed to accommodate the administrative structure of few states (Kerala and West Bengal). This was possible due to the adoption of accommodative and flexible approach in the SECC project.

The role of external stakeholders are important, though to a limited extend. However, the core and internal stakeholders gave due care to establish an effective communication channel to engage with them. General public, targeted public, welfare implementing agency and civil society organisations are the stakeholders external to the SECC project and through audio-visual advertisement and website information regarding the project was communicated. Welfare implementing agencies were made active party to the SECC project by assigning them the vital duty of selection, training and supervision etc. General public were also encouraged to participate in the claims and objection process. In this claims and objection stage, all individuals in both rural and urban areas were asked to go through the data that is collected (except caste information) and raise objections to incorrect facts and raise it in the Gram Sabha (village congregations) forum or as an independent individual and demand corrections.

Conclusions

Success of any project depends upon the role played by all the stakeholders. Knowing the nature of stakeholders and the stakes is the key to manage them properly to achieve the main goal of the project. The above sections show the complex role played by the stakeholders and the communication and engagement structure that was constructed and adhere to by the key stakeholders. In such a big and complex project of sociopolitical nature, predicting the nature of the various stakes and the behaviours of all stakeholders may not be perfected by each and every stakeholder. And the consequences of such imperfections result in cost escalation and time over run.

References

Sahu, N.K (2006). *Electoral Politics in Federal India: MP Local Area Development Scheme*. New Delhi: Gyan Publishing House.

6

N.K. SAHU

The Moving Target

Independent Monitoring of SECC by Third Party Agencies

Finding a target in physical, social and metaphysical space is not easy. We have three types of suitors in traditional Indian mythical ages. They are '*shabda vedi*', '*lakhya vedi*' and '*mana vedi*'. The one who can strike the target listing to the sound emanated from it, the one who can strike the target that is visually detectable and the one who can strike the target that is in the mind of the possessor of the target. In age of *Ramayana*, the king Dasharatha killed innocent Rishikumar (son of a hermit) by mistake as he managed to direct his arrow to a sound created by a pot while the Rishikumar was collecting water. In the *Mahabharata* age, Arjun the middle Pandava could strike the eye of a fish that was circulating on top of a pole looking at its image on the surface of water kept at its' bottom. Another warrior in *Mahabharata* age, Belarasen alias Barbarik could find the toe of Lord Krishna, though he hides his toe visually from him but the weapon read the mind of Lord Krishna. Now in present day warfare, the modern missiles are equivalents of the traditional arrows.

In modern day warfare, missiles are equivalent of bow and arrows of traditional times. The missiles may be cruise or ballistic, but the important parameters are the launch mode, range, propulsion, warhead and the guidance system. The last is probably the most crucial to find the target. The basis of guidance system are wire guidance, command guidance, terrain comparison guidance, terrestrial guidance, inertial guidance, beam rider guidance, laser guidance and radio frequency and GPS (global positioning system) reference (*http://www.brahmos.com/*). The advanced guidance system helps the missile to find mostly the target, only when the location of the target is known or the target moves in a particular speed and direction. Otherwise the probability of missing target increases. In this chapter, we will describe the designing of a 3rd party independent monitoring system of the Socio-economic and Caste Census (SECC) that is no less challenging in comparison either to the puranic warfare or the modern day's scientific warfare.

The Task

The credibility and reliability of outcomes of SECC is paramount to government. Millions of rupees, mostly the taxpayers' money, would be spent either as subsidy or assistance. Wrong beneficiary selection would not only be drains on the exchequer without achieving the intended results, but also detrimental to the credibility of the government's legitimacy. Thus a number of systems are to be put in place while conducting the SECC. People's participation and transparency are two such components of SECC. Besides other, allowing important locals to be present while enumeration is done resulted in enhanced credibility through peoples' participation. Similarly transparency was achieved by allowing displaying the data to public and open it for discussion and scrutiny in Gram Sabha—the village assembly. But this may not result in intended results as there is no guaranty that participation of public can be ensured. Experience tells that participation of public in Gram Sabha is below the required quorum as the democratic processes are not deep-rooted in many parts of the country. Many a time, the democratic process is subverted due to the local fragmented politics of powerful groups within local areas. To make up for these deficiencies, it was decided to put in place a third party concurrent monitoring system. The objective was to improve the quality of the outcomes of SECC—the household data.

The Target

There are five stages in SECC: Enumeration, Supervision, C&V (corrections & verifications), C&O (claims & objections) and Final list publication. These processes were structured and manuals were prepared and training was imparted to all the functionaries. TPCs (tablet personal computers) were utilised to collect household data and transferred it to the local servers for further quality improvement and onward transfer to central server. Though a number of checks and balances were put in place to maintain the quality, it was also thought to improve the quality by taking in to account the observations and suggestions of third party monitors called IMAs (independent monitoring agencies). As these IMAs are outsiders and professional research institutions, it is expected to not only give unbiased suggestions, but also enhance the credibility of the system. The question is to harness these unbiased suggestions when it would be useful and implementable. After study of all the processes involved and the time sequences, the most suitable time and locations of interventions were identified as below:

		1	*2*
A	Time of interventions	After enumeration is complete, but before the supervisory round starts	During the claims and objections stage
B	Location of interventions	Village/Urban area and enumeration blocks (EB) level	Village/Urban ward and Gram Panchayat level

A1 is the time when enumerators and data entry operators collect all household data in the assigned EBs (enumeration blocks) of a village or an urban ward and the data is loaded in the local servers at the community block offices. Next starts the supervisory round. A sub-set of households are to be supervised and data entered in TPCs and subsequently stored at the local servers. Once the enumeration and supervision is over and both data sets transmitted to the central server, no permission is given to change the data. And thus, any change is possible only when the full enumeration data is rejected based on the report that these are wrong entries or the supervisors came back with modified household data (a sub set of all household data). The decision either to reject the full enumeration data of one or more EBs or modify a sub-set of all the data depends on the reports collected by the higher officials involved in SECC and the apiary knowledge about the socioeconomic profiles of the targeted households. Besides, observations and suggestions of IMAs can also be the basis of arriving at such decisions. Thus the observations and suggestions of the IMAs are of value only when it is implementable, that is before the commencement of the supervisory round. Thus the concurrent monitoring system is to be so designed as to harness the services of IMAs only within a narrow window of time—between enumeration and supervisory round. This period is as narrow as a couple of weeks. Thus the IMAs are to visit sample villages and urban wards with trained manpower in sufficient quantity after enumeration is over, but before supervision starts and also communicate the observations and suggestions to appropriate officials so that the latter takes corrective measures.

B1 is the sample villages and urban wards where the IMAs visit to conduct concurrent monitoring and submit their observations and suggestions to appropriate officials. Based on such inputs, if the SECC officials so desire, improve the quality by asking the supervisors to take care of the observations and suggestions of the IMAs. The challenge for the IMAs is also to cover all the sample villages/wards so that their observations and suggestions are representative in nature.

A2 is a time when C&O process on. It has three stages: display of draft list containing all household data except the religion and caste, holding of Gram Sabha (village assembly) to scrutinise the socioeconomic data and claims/objection settlement system. All these three stages are to be completed within a prescribed time frame. The total period, depending on local conditions and the decisions of the state governments and union territory administrations may range between 45 days to 82 days. Thus it is crucial to harness the services of IMAs (observations/suggestions) only during these stages separately so that the actionable points can be implemented. That's why the IMAs are expected to visit three times during the C&O period and report submitted to local officials so that necessary actions may be taken at local grass roots level.

Thus from the analysis of the very nature of the SECC process, it becomes imperative to depute IMAs at critical times and to sample villages and urban wards. IMAs are required to visit the sample areas at four critical times: (i) Stage-I: after enumeration is over but before supervisory round starts, (ii) Stage-II (A): while the 'draft list' containing the household data is displayed for public viewing and scrutiny, (iii) Stage-II (B): during the holding of Gram Sabha (village assembly) where the household data is discussed for elicitation of facts, and (iv) Stage-II (C):during the time when claims and objections are settled by the disposing officer and appellate officer.

Stakeholders' Willingness and Capability

The four 'critical times' as mentioned in previous section depend on the willingness and capacity of the stakeholders not only to perform, but also to perform as per time schedule. Thus it is imperative to know the stakeholders' before the design of the concurrent monitoring through the independent agencies is finalised. The stakeholders' role and other relevant features are explained briefly below:

a) Ministry of Rural Development (MoRD) is the nodal ministry of the union government and has been assigned the task of organising the whole project. Finance is to be arranged and allotted to all other stakeholders. Budgetary provisioning are to be made during the life of the project. Prepare the rural questionnaire, field test it and get it approved at appropriate level. Build in-house capability by manning appropriately and set-up the central coordination team to overlook the whole SECC affairs. Impart training to the manpower that comprises the national level trainers. Besides, monitoring and supervising all the activities of the project so as to ensure quality of the outcomes and timelines of delivery of the results.
b) Ministry of Housing and Urban Poverty Alleviation (MoHUPA) monitors and supervise all the activities of the project in urban areas. Designing the urban questionnaires and imparting training to national level trainers is also done. They have a team at ministry level to provide the technical and supervisory services.
c) Office of the Registrar General and Census Commissioner of India (ORGI) is nodal for caste data collection. As they could

not collect this data during the general population census held in year 2011, this aspect is covered separately during the SECC. They handle the conceptual issues relating to 'caste' and train the trainers at the national level. Supervising and coordinating this operation through mainly the Directorate of Census Operation located at state head-quarters is done in a continuous manner and quality of caste-returns is ensured.

d) State governments and UT (union territory) administrations being the main stakeholders, organise the census at the ground level. The enumerators, supervisors, verifiers, charge officers, district/state nodal officers and much other manpower were selected, recruited, trained and put to service by them. Managing finance and manpower is their main responsibility. Storing and maintenance of TPCs made available by MoRD and putting these to use is its responsibility. Setting of 'charge centres'—an official space at community blocks and/or *Tehsil* offices with PCs, servers and power supply and other facilities and maintaining these are done. Security and quality of data is ensured.

e) Technical manpower and services, mostly to operate TPCs, charge centres and transfer data to central servers is assigned to a consortium of three CPSUs (central public sector units). Software needed for both TPCs and local servers maintained at charge centres are custom made by M/s Bharat Electronic Limited (BEL). Lakhs of DBAs (data entry operators) to operate the TPCs and accompany the enumerators and supervisors were recruited along with other technical manpower that were recruited and trained by CPSUs.

f) The central server where all the household data is stored and maintained is managed by National Informatics Centre (NIC). Hardware for storing the data and proving the required technical services was provided, of course with funding by MoRD.

Actually the numbers of stakeholders include 29 distinct states and 7 UTs. 39 of these are government bodies having mandate to provide public services, 3 are public sectors units that are commercial ventures and the last is NIC that combines public services with non-profit commercial activities. The structure and functioning of these agencies differs to great extent due to the very nature of these agencies. The resources at its disposal and the principles of incurring expenses and keeping accounts differ. Technical and managerial capability also differs. Thus delivering the expected services by these agencies may not be as per the agreed delivery schedule and quality. Thus it is imperative on the part of the MoRD to design the concurrent monitoring system of SECC in such a way that the objectives of gathering feedbacks at the ground level and facilitating the appropriate authority to take due actions so as to improve the quality of SECC outcomes is achieved in a transparent way. The deliverables expected out of the IMAs and the process of doing the concurrent monitoring should be such that the system is fair to both client and the service providers. Government–the client, should not loose financially but get the required services. The service providers, in this case the IMAs, should get the agreed price for agreed quantity of deliverables. But as we observed above, the very nature of movements of the SECC activities like the enumerations, supervisions and claims and objection are unpredictable as it is not similar with regard to the capability and willingness of the stakeholders.

The Key and Crucial Feature of the Concurrent Monitoring System

The standard features of a good monitoring system were adopted. Inputs, outputs, outcomes, goals of the system were studied and analysed. Based on this, the process features were identified and prioritised. The details on the stages and processes followed in this concurrent monitoring project with reporting timelines is explained in a summary tabular format in Annexure. But keeping the very nature of the stakeholders in view, an innovative clause was introduced. The relevant portion of the terms of reference is reproduced below:[1]

1. Ministry of Rural Development: Para-6, TERMS OF REFERENCE Concurrent Monitoring Socio-Economic and Caste Census 2011

> Period of Operation and Time Frame
>
> Monitoring reports are of any value only if it is actionable and are made available to appropriate authority at appropriate time. Though the period of operation for the concurrent monitoring of SECC would be of about 6 months, real action at EB and GP level would confine to only a few days. It would start concurrently with the launching of the SECC in a State/UT and end with completion of all the stages/ processes.
>
> However, the monitoring would be done in each GP in discrete stages, viz. enumeration stage (about 5 days), Information dissemination (about 3 days), Gram Sabha (about 3 days), claims/objection stage and appeal stage (about 10 days). The Independent Monitoring Agency (IMA) is to keep close contact with the authorities responsible for SECC at various levels and act promptly and in appropriate time to undertake various activities required for the concurrent monitoring. Waiting for instructions from MoRD or the State Governments should be avoided.
>
> As SECC was started in June 2011, it is in various stages of completion. Thus the Independent Monitoring Agency (IMA) should undertake the monitoring activities in sample GPs only for the stage(s) for which its reports/observations are of value and actionable. The IMA would take decision to do the monitoring work only after confirming the status of progress of the SECC work in the GP from the "Charge Officer"—the officer who is responsible at the Block/Tahsil level for the SECC 2011.

The above clause was the crucial feature of the concurrent monitoring system that ensured the deliverables in an environment of uncertainty. The uncertainty in SECC processes and the resultant outcomes became uncomforting reality and the concurrent monitoring system designed to be carried out by 3rd party independent monitors worked within this environment. The picture is made clear in Figure 6.1.

Figure 6.1 demonstrate that not only the starting and completion of the enumerations and supervision differs between states, but also the duration of period. When Lakshadweep took about 200 days, Punjab took more than 900 days. District level figures shows much more diversities. Thus foreseeing starting, ending and durations was next to impossible. However, the design of the concurrent evaluation system was such that the uncertainty associated was hedged and damage was minimised.

Figure 6.1

Durations of Enumeration and Supervision: State-wise

To ensure the field visit of IMAs in the critical window of time schedule (within a week or 10 days), it was insisted that the reports on each visit or stage should be submitted to the appropriate government official at the block/*tehsil*/*taluka* level. And the copy of the receipt with seal and signature in the prescribed format must be uploaded in the web-based monitoring site designed and made functional for this concurrent monitoring project. This web-based monitoring site (*http://secc.gov.in/wmt/Login*) is a restricted one and the officials and designated persons of IMAs can log-in for loading of data/files and monitor the progress.

Losses to Independent Monitoring System

Though the concurrent monitoring system was designed to hedge the uncertainty in the outcomes of SECC in presence of multiple of stakeholders and the vast and complex processes, loss to the whole system in terms of financial and physical achievements varies. Financial loss to government was nil and minimum to the IMAs. The IMAs were not allowed to gain financially without doing the job—though the amount of job done is much less than what is required for full coverage. They also did not incur financial losses as the design of the system allowed them to avoid underutilisation of resources, both human and material. The main looser was the system, as the coverage was much less than it is required to get a statistically significant result. Table 6.1 demonstrates this.

From the table, it can be inferred that out of 36 states/UTs targeted for coverage, only in 17 States the concurrent monitoring could be possible. Secondly, in these states, only in case of 4 states (Bihar, Mizoram, Uttar Pradesh and West Bengal), both the stages could be conducted. Thirdly, even though in these states, the stipulated sample GPs and urban wards could not be covered. At all India level, only 77.7 percentage of sample districts, 67.9 percentage of sample GPs and 62.5 percentage of sample urban wards were covered.

Table 6.1

Sample of Districts, Gram Panchayats (GPs) and Urban Wards Actually Covered by IMAs

S. No.	Name of State	Percentage of Sample Covered by IMAs			Comments
		Districts	GPs	Wards	
1	Andhra Pradesh	9.1	10.4		Only Stage-II
2	Arunachal Pradesh	100.0	100.0	100.0	Only Stage-I
3	Assam	100.0	104.7	100.0	Only Stage-II
4	Bihar	100.0	96.5	75.9	Both Stage-I & II
5	Gujarat	100.0	56.3	5.2	Only Stage-II
6	Haryana	100.0	48.9	48.6	Only Stage-II
7	Jharkhand	79.2	39.8	33.3	Only Stage-I
8	Karnataka	100.0	100.0	104.6	Only Stage-II
9	Kerala	92.2	89.5	81.0	Only Stage-II
10	Madhya Pradesh	36.0	52.0	34.6	Only Stage-II
11	Maharashtra	30.3	24.4	41.7	Only Stage-II
12	Meghalaya	85.7	80.0	75.0	Only Stage-II
13	Mizoram	100.0	100.0	100.0	Both Stage-I & II
14	Rajasthan	75.8	74.9	60.0	Only Stage-II
15	Uttar Pradesh	84.7	82.9	80.2	Both Stage-I & II
16	Uttarakhand	100.0	104.0	100.0	Only Stage-II
17	West Bengal	100.0	93.7	115.3	Both Stage-I & II
18	All India	77.7	67.9	62.5	

Conclusions

Firstly, the IMAs are private parties and thus their response to participate in the concurrent monitoring system was voluntary and based on financial viability. Thus, when the offer was to assign the task in all the states, the IMAs applied for only a few. They agreed to undertake the independent monitoring only in states of Andhra Pradesh, Arunachal Pradesh, Assam, Bihar, Chhattisgarh, Gujarat, Haryana, Jharkhand, Karnataka, Kerala, Madhya Pradesh, Maharashtra, Meghalaya, Mizoram, Odisha, Rajasthan, Tripura, Uttar Pradesh and Uttrakhand. But could not lunch the monitoring works in three states namely Chhattisgarh, Odisha and Tripura.

Secondly, in the states where the IMAs conducted the concurrent monitoring, both the stages could not be carried out. Thirdly, as it takes time to implement any new system, both the sponsor (MoRD) and client (IMAs) came to term with this concurrent monitoring only in the Stage II. This resulted in more close and systematic monitoring of the peoples' participation process of SECC.

In such a project like SECC, where the time schedule is subjected to actions or/and inactions of multiples of stakeholders, it becomes very difficult to engage and take works out of the IMAs that are private entities and not under the control of the sponsor—MoRD. The objective of the sponsor was not to incur financial loss, but achieve the aims and targets as far as possible. Thus the trade-off between financial prudency and achieving project objectives are in play in such a project. Thus, the dependency of the concurrent monitoring project on the main SECC project resulted in sub-optimal outcomes—less coverage, of course with efficient finance management.

7

SAVITA KULKARNI

A Step towards Gender Integration

Introduction

The term gender broadly refers to women's relative position in the society and its various systems. So, gender does not mean 'women' but women's status in relation to men and vice versa. The inequality in the society is based on various aspects like caste, religion, education, economic position, landholding—a world created around 'biological sex' of a person is an important factor which triggers discrimination in various sphere of life—personal as well as social.

Women have the equal right to participate in development of their communities and assessment of their needs. However, women 'miss out' from various programmes for various reasons. One, because of their invisibility in the social and economic arena; two, because they often don't voice their needs or they are not heard and third a general assumption that economic status is all about a 'family as a unit', thus ignoring intra-family disparities. This leads to an incomplete understanding of poverty.

In India, the status of women is assumed to be secondary to a great extent. Men generally dominate participation in government initiated programmes because women's mobility is limited, women are loaded by work and the environment is not always congenial for their participation. Women have shown a great leadership and community management skills through SHGs (self help groups) and other women centred programmes. However, when both women and men are expected to participate, generally women are pushed at the back, even they become invisible. When men provide information about women, it is filtered through their biases and hence not always authentic.

Knowing this ground reality, Socio-economic and Caste Census (SECC) team was keen to improve/increase women's participation by creating space and congenial environment for them. Looking at the scale of implementation (each household in the country was expected to be canvassed), it was important that nothing was left to chance and sensitive individuals, but create a gender sensitive design and implementation approach for the entire cycle of SECC 2011. The following were important stages of SECC: pilot study in 254 villages (August to November 2010), survey design, enumeration and supervision in field, software development, training, monitoring, claims and objections and data analysis.

At each of these stages, women's enhanced participation was aimed at and paid attention to, through various measures.

Gender Subgroup

During the Pilot Socio-economic Survey (PSES) stage, in November 2010, Ministry of Rural Development (MoRD) took an important decision of constituting a Gender Subgroup to integrate gender in SECC. At that time, it was called 'BPL Census'. This subgroup brought together senior policymakers, researchers and members of the larger advocacy network along with UN women. The members of the subgroup had experience of understanding the linkages between gender, poverty and deprivation. This group was established by an executive order from the Secretary MoRD and therefore inclusion of gender issues was institutionalised.

The Gender Subgroup was actively involved in PSES through inputs for survey design, observation visits to the field area and support for data analysis. The Gender Subgroup conducted number of meetings to review the SECC process from gendered perspective and also to give adequate inputs so as to enhance participation of women. The Gender Subgroup also provided inputs for media campaign. The Gender Subgroup provided various ideas to analyse SECC data.

As a member of the Gender Subgroup, UN Women provided a full time Gender Consultant to MoRD to support the process of gender integration.

Gender Sensitive Approach to Data Collection

In Indian communities, there are various stereotypes regarding roles and responsibilities of women and men. Women and men have differential access to resources. It is clear who can move in the 'public' sphere and who is responsible for 'private' sphere. Men have control over what women do in the private sphere. The patriarchal value system is inside the house and also influences outer activities. What women should do, can do, have to do, never should do...it is a long list of 'expectations' from women.

Box 7.1

SECC 2011: AT A GLANCE

- Women population: 48.16 % (rural), 47.79% (urban).
- Women headed families: 12.88% (rural), 13.69% (urban).
- Women with at least one disability: 0.63% (rural).
- Literate women: 27.99% (rural), 37.58% (urban).
- Women over the age of 65: 2.69% (rural), 2.59% (urban).
- Widow: 3.01% (rural), 3.12% (urban).

Enumerators (both women and men), with all their educational background and work experience are members of the same communities and hence are influenced by the same stereotypes. It was difficult to bring in considerable change in their perspective through a two or three days training programme. So, care was taken that gender sensitivity is built in the survey Schedule and during training the following points were elaborated to a certain extent. This approach generated interesting discussions amongst enumerators and helped to keep focus on enhanced participation of women. The points highlighted below are from the rural schedule, though some of the points are common to both rural and urban schedule.

a) *Declaration of separate household by women even when common kitchen was used:* SECC provided a unique opportunity to women. If any female member of a household decided to declare herself as a separate household, she was to be recorded as a separate household. Widowed, separated, second wives, single women are some examples of women who could declare themselves as a separate household. This was a breakthrough from conventional definition of household to reach out to poorest of poor women. This concept created strong debate and in the process it helped many to look at the reality from a different point of view. For example, some of the colleagues strongly felt that 'just to get separate BPL card, families would encourage women to declare separate status'. On the other hand, others expressed that vulnerable women would be discouraged to do so due to fear of social practices and respecting 'family' as a unit.

b) *Challenging the myth of Man is the Head of the household'*: A woman may not self-report that she is a head of a household because by social upbringing she is taught to be subservient to a male in all the circumstance. There have been instances in earlier censuses where women have reported minor male children as the head of their household in the absence of a male head of household who may have died, or divorced or migrated. Sometimes they would tell father-in-law or brother-in-law or father or brother as 'head of the household/ family' even when the household is managed by the woman. In such situation, it was important to underline the fact that 'woman is the head of the family'. The enumeration process created a space for 'woman' to declare herself as the 'head of the family'. It is observed that female headed households are generally poorer and lack access to information and social networks. So, the survey tried to capture these households so that they can be reached out later.

c) *Women's identity*: It was ensured that the name of the mother of each person was recorded along with name of the father. Regarding religion and caste records, enumerators were reminded that the wife's caste and or religion need not be the same as of her husband. Similarly, a non-PTG (primitive tribe group) man can marry a primitive tribe woman and the vice versa. One point of survey was to record 'monthly income of highest earning household member'. During the training, a special effort was taken to make trainers and enumerators aware that 'a woman could be the highest earning member of the household'. Similarly, emphasis was made on the point that 'a man might not be paying tax, but a woman could be paying tax'. If the man does not have salaried job, enumerators were not expected to assume that woman in the family too did not have salaried job. It is observed that in certain areas if we ask about children, due to linguistic connotations (*aapko kitane bachhe hai*?), sometimes only the sons are reported. Enumerators were specifically reminded to ask about 'daughters and sons' and not about children or '*bachhe*' or similar local words.

d) *Disability data*: Discussions were facilitated with trainers and enumerators regarding sensitive nature of this data. For example, the data collection process required to take into consideration the possibility of families hiding the disability of their unmarried daughters. Even disabilities related to old women are easily forgotten by young members. Hence individual data needed to be asked more carefully.

e) *Women respondents*: During various levels of training, the emphasis was on the 'sensitivity of enumerators to women respondents'. Due to existing gender stereotypes, many enumerators assumed 'man' to be the 'head of the family' and did not enumerate the household in the absence of man. The other assumptions are rural women, if they are illiterate won't know about land ownership and other such information. Women also may take more time to open up with strangers, especially when the enumerators were men.

f) *Marital status*: For women, declaring themselves married is very easy in our country. Their widowhood could be mostly understood from the outer indicators in many areas. Women also share their widowhood easily compared to their disputed or unstable marital status. Many deserted women and their families do not like to label the status as 'deserted' because they keep the hopes of reconciliation. Similarly, divorce is a legal process, which may take considerable time depending on the complexities involved into it. At village level, sometimes there are no clear boundaries between deserted, divorced and separated. Taking into consideration sensitive nature of this point, it was possible that women in these situations were not recorded by families. There was more chance of 'missing women' if this question was not sensitively enumerated. Enumerators were encouraged to use local terms (without changing the meaning) for deserted, separated and divorced for ease of conversation, though on record the same terms were used across India. (Deserted was included in separated as both indicated vulnerability of the woman). Sometimes there was a possibility of enumerator finding out a girl married prior to completion of age 18. In this case the legal status of the marriage was not to be disputed by the enumerators. Enumerators were specially instructed not to judge women based on their marital status. This was done considering that women are well equipped to understand consequences of their decisions and/or sometimes women have no choices, it indicated their vulnerability.

Media Campaign

SECC developed various kinds of material to share information with different stakeholders. A booklet giving brief information regarding SECC 2011 was brought up in November 2011. With UN Support, a TV spot was developed. Radio spots were developed by the MoRD. A set of six posters was developed and disseminated to the village level to provide information to people. The Media campaign material was utilised to gender sensitise the public. One poster appealed to women 'to provide information to enumerators, even when the man in the family is not present'. Another poster said that 'women can declare themselves as a separate household if they want'. It also said 'women can be the heads of the households'. Women were prominently portrayed in these posters and TV spots, thus encouraging them to participate in SECC .

Challenges and Strategies

SECC team explored the possible challenges in participation of women in the enumeration process. Some simple strategies were devised to address those. Table 7.1 elaborates some examples.

Table 7.1

Women Participation in the Enumeration Process: Challenges and Strategy

Challenge	*Strategy*
Generally, women don't come forward to answer enumerators, especially when the enumerator is a man and a stranger.	Enumerator should introduce herself/ himself and assure the woman that her answers will be noted by the enumerator faithfully as it is. Don't say "Oh, I will come when *Baba*/Brother/Man is at home.
Sometimes women do not understand Hindi or English words used by Enumerator.	Ask questions in a simple and local language. Avoid using English words. The questions should be asked as per the schedule and its meaning should not be changed.
Women get pressurised, due to either familial/societal dynamics or as enumerators are always in a hurry to finish the survey.	Don't hurry while asking questions. Be patient with the pauses, especially when women are providing the answers.
Women not being used to interact with outsiders, fear that they will give wrong answers and invite wrath of men in the family.	If a woman has answered the question, never ask the man if her answer is right or wrong. This will affect women's confidence.
When the understanding is that the survey outcomes would be utilised for beneficiary selection (BPL identification), a women may hesitate to act as the respondent, fearing non-selection may attract blame to her.	Enumerators were told not to give any impression that the survey is for beneficiary selection. Even, when such situations arise, this issue may be clarified in clear terms.

Achievements

a) *Gender sensitisation*: A large number of government officers and staff participated in the training programme. The design of the survey and the discussions during the sessions certainly helped these staff to think differently.

b) *Gender sensitive language*: A very simple but highly effective change could be immediately seen in the language used by SECC documents. Manuals, guidelines, instructions used he/she (instead of just he), himself/herself (instead of just himself), reminding people that both women and men have equal right to participate (or not to participate) in SECC .

c) *Access to technology*: Women staff using tablet for data collection was an amazing step in many ways. It clearly emphasised the fact that if women get opportunity, they can do any work. This was an empowering experience for many women enumerators.

d) *Disaggregated data*: Gender disaggregated data is available now for the entire country. The SECC data has certain limitations to portray complete picture of gender disaggregation. However, the data would certainly show intra-household disparities e.g. level of education, identification of widows etc. The data should be used for identifying women's needs; both practical and strategic; and to develop plans to address those. The SECC data would be able to highlight the types of work being done by women and also the capacity development they need to have a more meaningful livelihood. As the SECC is in the public domain, not only Central and State governments, but researchers and NGOs have the access to data to plan their initiatives to eliminate gender disparities in the country.

Conclusion

Creating enabling environment for women to participate in SECC data collection was in itself an achievement. It remains to be seen how the data is analysed with the perspective of understanding situation of women in reference to men.

Deprivations, vulnerabilities and poverty are a complex and multidimensional reality. Within the family, women are always the first who get affected by scarcity of resources. SECC has taken an important step in collecting some evidence in understanding the intra-family disparity. In a way, the SECC 2011 process has recognised the issue of identifying links between gender and deprivations/ vulnerability/poverty. There has been a conscious effort to capture women's vulnerability within the family.

Since 1992, Government of India conducted three censuses to assess poverty and identify poor families. SECC was a Census of similar objectives, but for the first time gender aspects were considered in the design and in the actual enumeration.

Some of the experts might feel that the process expressed here is very preliminary. Taking into consideration scale of implementation, time constraint, cultural diversity, heterogeneity of enumerators and the complex nature of the process, it is certain that an impressive beginning has been made.

Part 3

Technology Support System

8

SANTHI SOZHAN and TEJ TARUN MAHINDRA

Introduction of Tablet Personal Computer for Household Census in India

Introduction

At the time of data collection, a number of considerations are kept in mind. The main considerations are, nature of the respondents, volume of works, timeframe of the survey, criticality of the correctness of data/information, communication strategy between the interviewer/enumerator and respondents, and the resource (financial and non-financial) availability with the project. Face-to-face survey with the help of paper-and-pencil interviewing was the norms in the traditional surveys. But these days, due to advancement of science and technology, many new modes of data collection are adopted world over. The prominent are, telephonic survey (computer-assisted telephone interview, interactive voice response system, etc.), mail survey (post and telegraphic and e-mail) and web-assisted-survey. Socio-economic and Caste Census (SECC) was designed to cover a vast country called India—the seventh—largest country in the world, with a total area of 3,287,263 km^2 measuring 3,214 km from north to south and 2,933 km from east to west. Secondly, each and every household and its member are to be covered. Quality of data was the topmost priority as government assistance, both financial and non-financial would be targeted to the households/individuals selected from these respondent households based on objective norms. Thirdly, time frame to complete the survey was very-very narrow. Keeping these factors in view, it was decided to go for face-to-face survey with the help of TPC (tablet personal computer). The TPCs were used for the first time in the country for a massive data collection on SECC 2011 by Government of India (GoI). This entire exercise was done in an e-mode with the support of indigenous hardwares and customised software. And SECC 2011 is called the first paper less census done in the country, perhaps in the world. The different facets of introduction of TPC in SECC have involved various issues that need elaborations. How the entire task was managed by is presented here.

TPC: Markets Prior to 2011

There were various brands available during 2010-11, some of them are: Datawind, Samsung, iBall, Micromax, HCL-me, and some of Chinese brands etc. But the government owned Bharat Electronic Limited (BEL) Bangalore was requested to make TPCs for use in SECC.

Specifications of TPCs

While planning for acquiring large numbers of TPCs, many existing models were considered, Aakash being the one of them. In those days, the Ministry of Human Resource Development (MoHRD) was procuring a low-price basic TPC for school students being manufactured/assembled by DataWind. Indian Institute of Technology, Jodhpur (IIT-J) was consulted to arrive at the specifications of the proposed TPC keeping in view the requirements of the large scale survey (face-to-face interview). The basic requirements of TPC specifications such as 7 inch screen, 8 hours battery backup, solar charger was mentioned in the requisition order of Ministry of Rural Development. But in the end, the specifications advocated for the TPCs are given in Table 8.1.

Price and quantum of availability: The TPCs are differentiated products. Thus comparison is not possible strictly. However, the TPC provided by BEL, a GoI enterprise under the Ministry of Defense as per the required specifications and delivery schedule, cost at approximately ₹4,725. Other brands were also available at that time, but most of them were costlier as compared to BEL TPC. More over the quantity required was 6.4 lakh which was the biggest challenge on this front as none of the vendors/manufacturers were ready to meet this huge demand. Besides, the ability to deliver these many TPCs at different locations across the country posed additional logistic challenges. On this front the BEL has agreed to manufacture, transport and provide necessary ancillary items (solar battery-pack and solar chargers) along with TPCs and also rendering necessary supporting services.

Table 8.1

Specifications Advocated for the TPCs

No	*Description*	*Specifications of TPC without External Keyboard and SD Memory Card*
Hardware		
1	Processor	450 MHz
2	Memory (RAM)	256 MB
3	Storage (Internal)	2 GB
4	Storage (External)	SD card (2GB up to 32 GB)
5	Peripherals	USB 2.0 ports (2 nos.)
6	Audio out	3.5 mm jack or integrated speakers
7	Audio in	3.5 mm jack for external headphones/speakers or integrated MIC
8	Display	7”
9	Resolution	At least 800x480
10	Input devices	External Key Board
11	Connectivity and networking	Ethernet 10/100 Mbps (RJ45)
12	Power	4 Hours 4000 mAH
	Power adaptor	5V, 1.5 A (Additional battery of 4000 mAH for additional 4 hour backup)
13	Operating system and system software	Android Version 2.1 or above
14	Document rendering formats	DOC, DOCX, PPT, PPTX, XLS, XLSX, ODT, ODP, ODS PDF viewer Text editor
15	Multimedia and image display	Image viewer to support PNG, JPG, BMP, Tiff and GIF display
16	Audio	MP3, AAC, AC3, WAV, WMA
17	Video	MPEG2,MPEG4 (AVI container, available through ffmpeg), FLV (Flash lite version 3 or later)
18	Communication and internet	Web browser (xHTML 1.1 compliant, JavaScript 1.8 compliant) Optional Flash 9 or later support (through plugin) Optional Java support (through plugin)
19	RoHS certification	Compliant for RoHS certification

TPC Manufacturing Facilities in India and in World Prior to 2011

According to research firm IDC, about 250,000 tablets were sold in India in calendar year 2011, which is expected to increase to 1.74 million in 2012. The government's appetite for low-cost tablets gave birth to its manufacturing businesses. For example the Wish Tel, which started in surveillance systems (like security cameras), entered tablet manufacturing in December 2011 with the specific intention of catering to government demand. The current international assessments visible from various sources add up to 10 million tablets, according to numbers released by hardware body Manufacturers Association of Information Technology (MAIT). Although it is not clear over what period this offtake will materialise, this number is about five times the total sales that research firm IDC expects in India in calendar 2012. In the year 2011, the Ministry of Rural Development (MoRD) bought 600,000 solar-powered tablets from Bharat Electronics at $72 (₹3,600) per unit for its staff to conduct the SECC. In the pipeline is a jumbo-sized order of 5.86 million tablets from the MoHRD, which it plans to give out at a subsidised price of $35 (₹1,750). The first phase of this is for 1,00,000 units of the Aakash tablet, the second version of which is due to be released by Canadian manufacturer Datawind.

Planning to Procure and Manufacture

The BEL has grown into a multi-product, multi-technology, multi-unit company servicing the needs of customers in diverse fields in India and abroad. BEL is among an elite group of public sector undertakings which have been conferred the Navratna status by the GoI. Its experience in manufacturing for defence sector includes communication equipments, receiving valves, germanium semiconductors, radio transmitters, radar manufacturing, transmitting tubes, silicon devices and integrated circuits, frigate radars for the Navy, image converter and image intensifier tubes, magnesium manganese dioxide batteries, tank electronics, switching equipment and many other products. For civilian sector, it manufacture black & white TV picture tube, X-ray tube and microwave tubes, integrated circuits and hybrid micro circuits, TV transmitters for Doordarshan—the autonomous public service telecaster of India, low power TV transmitters and TVROs, TV glass shell and electronic voting machines. BEL has world class mass manufacturing facilities which includes SMT assembly, soldering reflow, X-Ray, microscopic inspection and automated machines like test jigs to test the boards during PCB level and highly skilled workers to test the PCB/product and labs are complied with the National Accreditation Board for Testing and Calibration Laboratories (NABL) norms.

Thus there was no dearth of manufacturing, assembling and product developing experience in BEL. For manufacturing of TPCs, like the case with the listed items above, experienced engineers were involved during the process of planning, procurement and manufacturing as per the laid down management principles followed by BEL. The top and middle level management personnel of BEL were involved in the decision-making process. Brainstorming sessions and practical experience of previous projects had helped in the decision-making process and appropriate decision were taken during each stage of manufacturing.

Designing the TPC

The MoRD has called for a low cost device with basic functionality of TPC with seven inch LCD with touch screen, Ethernet, USB, detachable keypad, additional battery pack and solar charger. On an average, the range of costing is between ₹3,000 to ₹4,000 per TPC. The operators/users engaged to carry-out SECC were the skilled as well as semi-skilled operators. The latest technology components were used in the design of TPCs. Skilled and trained manpower of BEL were used in the design and manufacturing. TPC with keyboard was like a book for easy handling and easy to carry, at the same time it can be used as laptop also because of the keyboard provided with it. In order to achieve aesthetic and ergonomic requirements, small form factor was considered. Some of the places where electricity is the main bottleneck, solar chargers were provided for smooth and seamless execution of the work.

Sourcing of Components

Indian and imported components are sourced for the TPC. Plastics, adaptor, solar charger and few passive and connectors were sourced from Indian manufacturers through e-tendering. Passive, active and assembly components were sourced from global manufacturers and sourced through e-tendering.

Components such as ICs, LCD, TOUCH etc. were imported by BEL from other country like Japan, Singapore, Taiwan etc. Due to lack of IC manufacturing facilities in India, it was the sole option for BEL to import from the foreign countries. The share of imported components is about 70 per cent of all the components that were used in the production of TPC.

Manufacturing/Assembling Process

The existing production line in BEL was utilised for the production of TPCs up to 1,500 PCB (printed circuit board) per day. New SMT (surface mount technology) line was installed to gear up the production of TPCs up to 1,500 PCBs per day. The procurement of new SMT line and installation within three months was really a difficult task to complete. These challenges were handled by using risk planning and mitigation techniques.

SMT pick and place machine is the machine used for the placement of the surface mount devices over PCB which is used to make compact or low/small form factor of TPC.

BEL uses two SIPLACE SX2 SMT machine having the following features:

- The two machines have the mounting capacities of 46,000 and 60,000 components placement per hour (max); and
- It supports components ranging from 0105 to 50x40 mm size and SOIC (small outline integrated circuit), BGA (ball grid array) and LGA (land grin array) packaging.

EM SBU had one SMT Pick & Place line and one extra line was added to cater the TPC production requirement. Both SMT lines were running 24X7 during the mass production of TPC. It took only five months to produce the 6,40,000 TPCs with the production capacity of 4,500 TPC per day.

Figure 8.1

Basic Flow/Process Chart for SMT Line

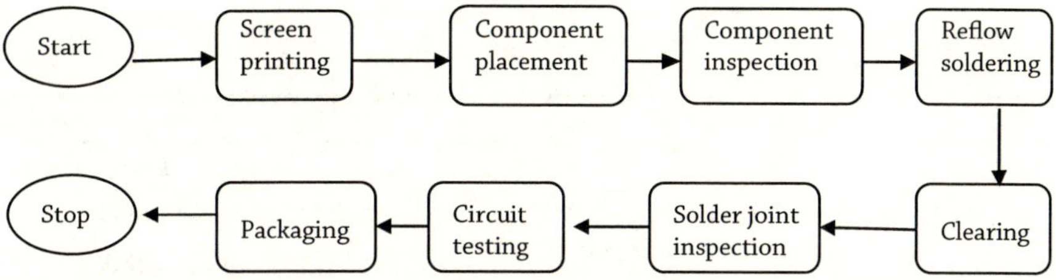

Manpower Sourcing and Training/Deploying

Existing manpower was utilised for manufacturing with additional manpower from various departments of BEL. Training was provided for the existing and new manpower to carry out the manufacturing activities in a shorter period of three to six months.

Although EM SBU/BEL has adequate production manpower, some skilled manpower who had the experience with SMT assembly, soldering reflow, testing etc. were taken from all the SBU across BEL during TPC production.

Quality Standards and Testing

Old labs were used to test existing products. Existing environment and various labs were utilised for the testing of TPCs. New labs were not established, but all the existing labs were strengthened with latest required equipments.

The following types of tests were done on TPCs:

a) All the tests related to the ESS (environmental stress screening) were complied with International Electrotechnical Commission-IEC68 standards and all the preventive measures had been taken during the tests.

b) Thermal shock and vibration test were conducted for each PCB.

c) Burn-in test of TPC was carried out in power ON condition at + 50°C for 8 hours for 100 per cent unit.

d) ESS qualification test specification details:

(i) EMI/EMC test was carried out by design group as part of product approval testing. The test specification are as follows:

Emission Test

Test	*Standard*	*Specifications*		
		Freq	*Dist*	*Field Strength*
Radiated Emission	FCC Part 15 Class B consumer computing device	30-88 MHz	3 mt	100uV/m
		88-216 MHz	3 mt	150uV/m
		216-1000 MHz	3 mt	200uV/m

SI. No.	*Test*	*Specification*	*Duration*
1.	Vibration test sinusoidal IEC 68-2-6	Freq. 5 to 500Hz ±6mm or 2g, which ever is less (10 sweeps)	10 minute each in X, Y, Z, axis.
2.	Vibration test random IEC 68-2-34	Freq. 20 to 2000 Hz 0.05g2 / Hz.	9 minute each in X, Y, Z axis.
3.	High temperature (Operational) IEC 68-2-2	+55°C	16 hours
4.	High temperature (storage) IEC 68-2-2	+ 70°C	16 hours
5.	Damp heat IEC 68-2-28	+ 40°C., 95/ RH	16 hours
6.	Low temperature (operational) IEC 68-2-1	0°C	16 hours
7.	Low temperature (storage) IEC 68-2-1	-30°C	16 hours
8.	Bump test IEC 68-2-29	40g 6m sec	4000 bumps

Immunity Tests

Test	*Standard*	*Applicable Limits*
Radiated RF electromagnetic immunity test	IEC61000–4–3 Level 2 moderate electromagnetic radiation environment. Low power transceivers (Typically < 1W). A typical commercial environment	Test field strength 3 V/m
Electro static discharge immunity test	IEC61000–4–2 Level 2 Relative humidity as low as 10% and antistatic material	Test voltage ±4KV

Electrical fast transients (Burst Immunity Test)	IEC61000–4 – 4 Level 1 On I/O signal data and control ports. Protected environment. Representative—Control room or terminal room of industrial and electrical plants	Voltage Peak 500V Repetition Rate 5 KHz

The various stages of tests involved are: PCB testing and Unit level testing.

SECC: Pilot

A pilot was conducted at Sonipat district of Haryana adjacent to Delhi for testing the performance of TPC and software developed for data collection. Prior to pilot, a training was given to the data entry operator who operated the TPC and enumerators were trained to ask the respondent with relevant questionnaire. A charge centre was established at Sonipat to facilitate the specific area with all the required facilities. TPCs were tested for: (i) backup of internal battery; (ii) backup of external battery; (iii) readability; (iv) ruggedness; and (iv) use of keyboard with TPC.

There were some issues like readability in sunlight which was resolved by increase in the brightness of the display. First prototype's casing was not rugged enough to manage the shocks on accidental drop of TPC , these casing was upgraded after pilot.

SECC software were also tested from installation to export of the data from TPC through enumeration of data in field.

Packaging and Delivery to Destinations

At the planning stage, it was known that the packing box shall have to contain the TPC, adaptor and user manual. Separate packing cartons for additional battery pack and solar charger was arranged. Packing design was carried out to accommodate the TPC and accessories; card board used as packing material. Separate packing box were used for TPC, solar charger and additional battery pack.

All the packaging related to TPC and TPC modules were carried out in the EM SBU only. Some automatic machines like automatic fitter machine and ultrasonic welding machines were used for the packaging of TPC additional battery pack.

Service of TPCs

There were various types of services provided for TPC, broadly the services provided for three types of problem: (i) replacement of broken keyboard; (ii) replacement of broken screen; and (iii) missing SD card.

State-wise Details on Quantities of TPCs Supplied

BEL supplied 6.4 lakh TPC to 33 states/UTs of the country through various mode like by road/air. Some of the states requested to deliver the TPC at State HQ and other requested to deliver at district HQ.

Once TPC are delivered at all the desired location, next steps is to store all the TPC at safe and secure place so that it can be easily shifted to corresponding charge centres for further installation of software and required files for field usage. List of TPC distribution are given in Table 8.3.

Conclusion

Due to the conduct of SECC project by MoRD, the biggest ever order of 6.4 lakhs of TPCs were met by BEL at short notice. The long standing experiences of BEL to meet both military and civilian needs and availability of functional manufacturing infrastructure and human resource proved to be very useful and handy while manufacturing such a huge order. The various challenges that came by in the whole processes were resolved in very systematic and innovative manner is evident from the above presentation.

Table 8.2

Distribution of TPCs among Different States/UTs of India

No.	*Name of State/UT (No. of TPCs)*	*Locations of Delivery of TPCs and Numbers*				
1	Andaman & Nicobar Islands (400)	South Andaman (400)				
2	Andhra Pradesh (60580)	Gollapudi (9294)	Hyderabad (140)	Kadapa (12242)	Pendurthi (9426)	Rangareddy (29478)
3	Assam (22960)	Guwahati (22960)				
4	Bihar (33040)	Patna (33040)				
5	Chandigarh (650)	Chandigarh (650)				
6	Chhattisgarh (15860)	Bijapur (240)	Bilaspur (1400)	Dantewada (320)	Dhamtari (440)	
		Durg (1720)	Kanker (440)	Koriya (400)	Narayanpur (160)	Raipur (9900)
		Rajnandgaon (840)				
7	Dadra & Nagar Haveli (225)	Silvassa (225)				
8	Daman & Diu (185)	Daman & Diu (185)				
9	Goa (1000)	North Goa (1000)				
10	Gujarat (36400)	Ahmedabad (520)	Gandhi-nagar (35880)			
11	Haryana (14000)	Ambala (3160)	Gurgaon (3000)	Hisar (4120)	Rohtak (3720)	
12	Himachal Pradesh (10360)	Bilaspur (400)	Chamba (672)	Gurgaon (1680)	Hamirpur (600)	Kangra (1520)
		Kinnaur (363)	Kullu (320)	Lahul & Spiti (265)	Mandi (1240)	Shimla (1420)
		Sirmour (480)	Solan (960)	Una (440)		
13	Jammu & Kashmir (8440)	Jammu (3960)	Srinagar (4480)			
14	Jharkhand (22040)	Ranchi (22040)				

15	Karnataka (40775)	Bagalkot (1040)	Bangalore (6550)	Belgaum (2655)	Bellary (1375)	Bidar (1000)
		Bijapur (1200)	Chamarajanagar (680)	Chickaballapur (1040)	Chickmagalur (880)	Chikaballapur (0)
		Chitradurga (1160)	Davanagere (1175)	Dharwad (495)	Gadag (600)	Gulbarga (1495)
		Hassan (1800)	Haveri (920)	Hubli (520)	Kodagu (400)	Kolar (1200)
		Koppal (800)	Mandya (1320)	Mangalore (1295)	Mysore (1935)	Raichur (1120)
		Ramanagar (1000)	Shimoga (2240)	Tumkur (2400)	Udupi (720)	Uttara Kannada (1080)
		Yadgir (680)				
16	Kerala (21970)	Alappuzha (400)	Ernakulam (400)	Idukki (400)	Kannur (400)	Kasargod (40)
		Kochi (50)	Kollam (400)	Kottayam (400)	Kozhikode (400)	Malapuram (400)
		Palakkad (17480)	Pathannamthitta (400)	Thiruvanthapuram (400)	Thrissur (400)	
17	Madhya Pradesh (50360)	Bhind (1200)	Bhopal (28720)	Chhatarpur (1200)	Datia (520)	Gwalior (3160)
		Indore (5720)	Jabalpur (3680)	Jhalawar (80)	Morena (1320)	Panna (480)
		Sagar (2120)	Sheopur (360)	Shivpuri (960)	Tikamgarh (840)	
18	Maharashtra (45910)	Ahmednagar (2320)	Akola (1000)	Amravati (1640)	Beed (1320)	Bhandara (680)
		Buldhana (1360)	Chandrapur (1280)	Dhule (960)	Ghadchiroli (760)	Gondia (680)
		Jalgaon (2160)	Mumbai (7640)	Nagpur (3800)		Nandurbar (800)
		Nashik (3120)	Navi Mumbai (700)	Pune (4930)	Thane (6320)	Thane (West) (1320)
		Wardha (920)	Washim (640)	Yavatmal (1560)		
19	Manipur (2120)	Imphal (2120)				
20	Meghalaya (2760)	Baghmara (280)	Jowai (280)	Nangpoh (240)	Nongstoin (400)	Shillong (560)
		Tura (1000)				

21	Mizoram (700)	Aizawal (700)				
22	Nagaland (1330)	Dimapur (1300)	Kohima (30)	Narayanpur (0)		
23	New Delhi (10830)	Delhi (10830)				
24	Odisha (31100)	Bhubaneswar (31100)				
25	Puducherry (700)	Puducherry (700)				
26	Punjab (16800)	Sas Nagar (16700)	Sas Nagar-Mohal (100)			
27	Rajasthan (44210)	Ajmer (1640)	Alwar (2240)	Banswara (1080)	Baran (840)	Barmer (1600)
		Bharatpur (1560)	Bhilwara (1520)	Bikaner (1520)	Bundi (720)	Chittorgarh (1080)
		Churu (1260)	Dausa (1000)	Dholpur (760)	Dungarpur (860)	Hanuman-garh (1400)
		Jaipur (4330)	Jaisalmer (480)	Jalore (1120)		Jhalawar (920)
		Junjhunu (1360)	Jodhpur (2360)	Kankroli (920)	Kota (1320)	Nagaur (1960)
		Pali (1400)	Pratapgarh (580)	Rajasmund (720)	Sawai-Madhopur (880)	Sikar (1600)
		Sirohi (620)	Sriganga-nagar (1680)	Tonk (960)	Udaipur (1920)	
28	Sikkim (520)	Gangtok (520)				
29	Tamil Nadu (43680)	Chennai (43680)				
30	Tripura (2250)	Agarthala (2250)				
31	Uttar Pradesh (54765)	Allahabad (10680)	Gorakhpur (12520)	Kanpur (12200)	Lucknow (8200)	Meerut (11165)
32	Uttarakhand (8960)	Uttar Kashi (280)	Uttara khand (8680)			
33	West Bengal (34120)	Burdwan (7520)	Darjeeling (2080)		Kolkata (21520)	Malda (3000)

9

TEJ TARUN MAHINDRA

Software Development for SECC

The Challenge of Meeting Multiple Demands

Introduction

The earlier caste census was performed in 1931 and that was a paper based census only. The Socio-economic and Caste Census (SECC) was conducted as supplementary to 2011 Census of India. Government approved the SECC 2011 based on the discussions in both the houses of Parliament in 2010. The SECC 2011 has three census components which were conducted by three separate authorities, under the overall coordination of Ministry of Rural Development (MoRD) in the Government of India (GoI). These are: Census in rural area conducted by the MoRD; Census in urban areas done under the administrative jurisdiction of the Ministry of Housing and Urban Poverty Alleviation; and Caste census under the administrative control of Ministry of Home Affairs (Registrar General and Census Commissioner of India). Once it was decided by government to conduct the SECC in 2011 using latest electronic technology, the next steps were to formulate the comprehensive system to execute the census. Unlike other national schemes, SECC facilitated the option of collecting data in two formats, one for rural and the other for urban areas. Separate methodologies were designed for urban and rural by Prof Hashim committee and Dr N.C. Saxena committee respectively. The different methodologies adopted were required to be converted and implemented at software level for field survey. Thus the software development needs for SECC 2011 has posed multiple challenges and demands and these are being illustrated herein.

Scope and Objectives

As it was the first of its kind, it was really a very tough task to anticipate the type of challenges that may come in due course of processes at the time of execution. SECC is the first of its kind of government project where in such a big canvas of technology was used for data collection as mentioned earlier. Catering to software requirement of SECC was challenging as demands of three core clients (MoRD, MoHUPA, and Office of the Registrar General & Census Commissioner, India [ORGI]) and 30 plus states are to be addressed. The software developed and involved for SECC had to perform few critical tasks. Broadly two types of software were used in SECC, first type of softwares were used in data collection in field, while the other one related to monitoring the progress of all activities. Android operating system was the only option available during 2010 when SECC was conceived for the execution of software on TPC (tablet personal computer) or any other HHD (handheld device) unlike present day scenario of operating system on HHD like Windows. Thus to work with limited options of using Android was also challenging.

During initial stage of SECC, only some of the processes were identified like data collection, and draft and final list publication. However, subsequently a pilot was executed at Sonipat, Haryana for data collection module and supervision modules through TPCs only, but draft list was not tested during pilot. How 'claims and objections' process would be handled was not conceptualised during that time, and hence, 'claims and objection' handling process had to be introduced later. In the process of data collection, some of the above observations were taken care of and affected at the central server. There was no provision to view or monitor the quality of data before it was uploaded on central server. It was decided in late 2012 to introduce DVM (data view module) so that the incharge officer of charge centres can see the data and flag/mark them for doubtful or/ and wrong entries so that these records collected on TPCs/HHD can be rectified during supervision stage.

Softwares Developed for Field Data Collection

First time in the history of use of electronic device, TPCs were used to collect the data of approximately 120 crore population of India. Once it was decided to use TPC, the next question came up was about the software to be used to manage the TPC at field level, covering lakhs of villages and urban wards. Software was planned broadly for two sections; one Android for TPC and the other one for Linux Centos for server/client machines deployed at charge centres. The charge centres are intermediate level administrative headquarters located between districts and villages and urban wards. Mostly these are blocks and/or *tehsil* offices. Both these operating system are free for use without any license fee, and this way, some of the taxpayer's money was saved. M/s BEL was assigned the task of developing various types of software to be used during the process of SECC.

Following flow diagram explains the type of processes involved in SECC and their sequences. Accordingly corresponding software setup is also required for these processes.

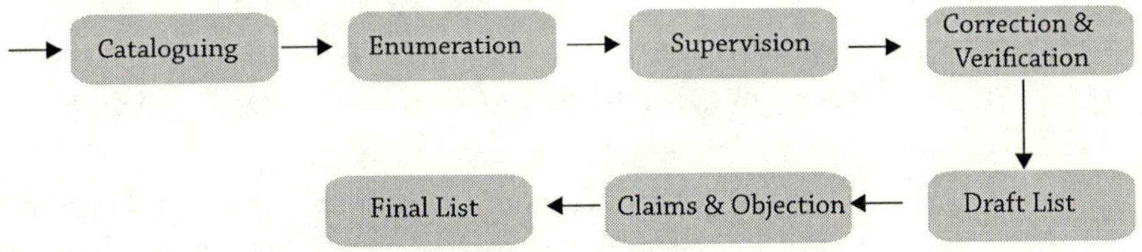

Cataloguing process (TIFF images to JPG images): National Population Register (NPR) images were taken as the base for the field survey and they were loaded in the TPCs. But during 2011, TPCs were manufactured at low specification as compared to today's specification of TPCs. NPR images (contains 25 schedules i.e. approximately 25 households data in TIFF format) cannot be handled smoothly on the TPC due to its formatting and size. The specification of TPCs were not sufficient to manage it appropriately, hence it was decided to segregate these 25 schedules from NPR TIFF images into JPEG format which is very small size in comparison to NPR TIFF images. Approximately 25 crore households are there in SECC and extraction of these households images

before commencing the process was required which in turns needed a software with manual intervention to read the AHL (abridged household listing) booklet provided by the ORGI along the NPR images. Once these images were converted to JPG images and transferred to TPCs, these tablets are ready for field survey. This whole process was carried out at charge centres where servers were installed by CPSUs (central public sector units) which were equipped with required software.
The charge centres too were planned in such a way that it sufficiently catered the requirement of cataloguing and proper charging facility for TPCs were given at the centres.

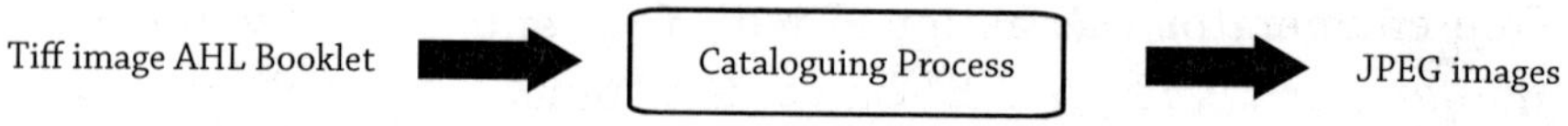

Enumeration: At charge centres, catalogued images along with various other prerequisites were installed in TPCs before sending them into field. A team of enumerator and DEO (data entry operator) was designated to each EB (enumeration block) to collect the data as per the JPG images converted in the above mentioned process. Enumerator and DEO were given training on respective subject for field survey. In such a survey, the enumerators and DEOs visit households in each EBs with TPCs where fresh enumeration on some basic pre-loaded information/data was to be done. They encountered the following situations:

- Household not available at the location: Moved inside (MI);
- Household shifted to some other enumeration block: Moved outside (MO); and
- Household might have been split.

To cater to above mentioned situations, the software had the provision to handle them, such as:

- Add new member reported during survey;
- New household can be added during survey also; and

- Current status of household can be updated like household shifted to some other area or a household divided into more households.

Once the field data collection is complete, the TPCs needs to be handed over at charge centre where data backup of that EB is to be done at server system. Subsequently in these servers, E-files (database file) were generated which needs to be upload on National Informatics Centre (NIC) website for central server. The central server is managed by NIC—a Government of India agency to provide various types on e-services. After upload of e-file to NIC server, the TPCs is again configured for next enumeration block for next round of field work. One enumerator was assigned with four enumeration blocks. As this survey was respondent based, the enumerator cannot question about the genuineness of the respondent, but he can provide remarks during data entry so that the supervisor can verify those facts and do the needful action during the supervisory stage. The SECC has provisions for unique location specific ID for each households. Such as, state, districts, block, town, villages, enumeration block, household member etc. These IDs have 29 digits and is called TIN (temporary identification number). TIN has following constituents:

XXSSDDTTTVVVVWWWWBBBBsHHHHMMM

1. XX: Generated as per the location in TIFF image
2. SS: State Code
3. DD: District Code
4. TTT: Tehsil Code
5. VVVV: Village or Town Code
6. WWWW: Ward Code
7. BBBB: Enumeration Block Code
8. s: Sub Block Code
9. HHHH: Household Number
10. MMM: Member ID

During this process, some of the states came with special requirement like Kerala and West Bengal to include SSLID (state specific location Id) and hence during the enumeration process the software was modified for these states. Each and every household have been provided with the acknowledgement slip after data collection. The specimen is given below, that necessitate specific treatment in software development and executions.

SOCIO-ECONOMIC AND CASTE CENSUS 2011
ACKNOWLEDGEMENT SLIP BOOKLET

Book No. (Pre Print)

Serial No.: 001 - 150

Identification particulars:

(1) Name of State/UT ________ Code □□ (2) Name of the District ________ Code □□

(3) Name of the Tahsil/Taluk/PS /Dev. Block/Circle/Mandal etc. ____ Code □□□ (4) Name of Town/Village ________ Code □□□□

(5) Ward Code (only for town) Code □□□□ (6) Houselisting Block No Code □□□□

(7) EB and Sub-Block No Code □□□□ □

Name of Enumerator ________

Signature Date

Name of Data Entry Opreator ________ Code

Signature Date

Name of Supervisor

Signature ________ Date ________

As mentioned above, the field level needs prompted redesigning and re-writing of the client's as well as server software many a times. Various modifications were done in data entry software along with server software. Latest client version in field was Cv0.64 and server version was Sv0.64, meaning, the client and server versions went through 64 sequential modifications respectively.

Data View Module: To facilitate the supervisory functions, the DVM was introduced so that the in the DCC (data charge centre) the authorised officer can select the doubtful variables/data that can be verified during supervision. The DVM enables the incharge of DCC to view all the data collected by the enumerators. This facility enables the charge officer to assess the quality of data collected. Wherever the correctness and authenticity of data collected is doubtful, such households (e.g. large number of legally released bonded labours than the official figures with the administration, religion and caste information misleading etc.) can be brought under the purview of the scrutiny by the Supervisor. Those households that were flagged by the charge officer for supervisory field check, can become part of the e-files for supervision. Only such files can be edited by the supervisor and remarks entered in the relevant column. The supervisor was to be assisted by a DEO to take the house to house visits.

Supervision: This module takes the identified/flagged data from DVM and random 10 per cent records for supervisory checking in field. Following are the types of households selected for mandatory supervisory verification:

(i) all un-enumerated households along with 'reason for non-enumeration':
  - all un-enumerated catalogued households image files as per AHL Section 2,3,4 of catalogued households.
  - Section 2 contains all the household found during NPR.
  - Section 3 contains new households.
  - Section 4 contains houseless households.

(ii) all households where the enumerator disagrees with the information furnished by the respondent;

(iii) caste/religion information missing in the relevant boxes;

(iv) the household's whose data is selected via the use of 'DVM'; and

(v) lastly if all the above types of households do not add up to 10 per cent of the total, then the deficit only will be made up from among the good records.

Once supervision round is complete, software generates S files which also needs to be uploaded to central server from charge centre.

Correction & Verification: This is a stage that was not conceived at the time of start of SECC. The V&C (verification and correction) Module was introduced to ensure the quality of data with respect to a specific set of parameters. The verification officers are distinctly different from supervisors, they will have to physically verify the mandated fields namely; the PTG (primitive tribal groups)/LRBL (legally released bonded labourer)/MS (manual scavenger) in the rural areas and for the urban areas the four types of major sources of income from rag picking, begging, domestic jobs and street venders. A select list of common errors in reporting the caste names/codes have been identified by ORGI and put to verification and correction. The files for verification are in PDF format and with relevant space obtaining the signature of the respondent and for writing the remarks of the verifying officer. An additional caste correction module also was introduced to take care of errors noted even after execution of V&C module.

(i) The need of V&C module has become imperative for the following reasons. From almost all the States and UTs the enumerated data have thrown up the following inconsistencies. First, a disproportionally very high number of households reported to have enumerated as members belonging to PTG, LRBL and MS. Secondly, in urban areas, data collected on the numbers of occupational categories like beggars and rag-pickers are found to be unreasonably high. Thirdly, an apparently wrong/ incomplete returns of caste/tribe name, e.g., 'other'/'open'/'general'/'OBC'/ abbreviated caste name/ religion name, etc. have been found in enumerated data.

(ii) In the implementation of V&C module, all the States/ UTs are advised to implement this module before draft list publication. With respect to the above five mandated fields, the household numbers with member details, EB numbers with their reported/enumerated data would be provided at the DCCs with the help of application software designed by NIC and BEL. There will be printed form for each of these households with earlier entry and space for new/verified entry. The State government will send an official designated 'verifying officer' to those particular households and get the correct data from the respondents.

(iii) The corrected data will be entered and uploaded at the DCC with the help of the application software and will be retained in the NIC server. The CPSUs will circulate the instruction manuals for the application software and organise training for the DEAs (data entry administrators).The service provider CPSU would ensure the engagement of DEAs in required quantity and quality.

Draft list generation for public display of data: Once above mentioned steps completed, data was ready for public display and claims. As JPG images were used during survey to identify the households based on the head/member of household by verifying fields like name, date of birth, father name, mother name, occupation and marital status that were not digitised during survey, cropped (JPG) images were used to generate draft list. But later on it was identified that different patterns of handwriting (in cropped NPR JPG images) was difficult to read. These fields were not digitised during the survey as these fields were already part of NPR project and it was planned to collect these fields from NPR project. Different versions of draft list software were released because of various requirements from State administration like from Kerala and West Bengal regarding SSLID issue. One more situation raised as in Assam different language for different districts were used as some district required draft list in Assamese while other need it in Bengali. Similarly, Bihar decided to handover the hard copy of the draft list to each household and hence it was planned to start printing every household from a new page. Thus for draft list generation, many versions of software were introduced to meet the local state specific needs. Latest draft list modules used were 4.0.5a, 4.0.6 and 4.0.7 by different states.

Claims & Objection Tracking System (COTS): This module for handling the claims was also not planned during initial stage of SECC. Six types of different forms were designed to cater the claims process. Respondent need to fill the corresponding form as per his/her requirement and submit at designated office.

Each and every claim received during claims & objection phase were assigned a unique number for tracking the status online. Latest version of COTS used in field was 1.5.17. Following are the various forms to cater the requirement of public:

(i) Form A is used for filling objections against inclusion of particular data.
(ii) Form B for making corrections and modifications.
(iii) Form C for inclusion of data.
(iv) Form D for issue of Notice to the Objector/Respondent.
(v) Form E for recording the proceedings of the Grama Sabha (not used for urban areas).
(vi) Form F for marking the respective corrections done at the time of disposal of claims. A duly filled and signed copy of Form F would have to be given to the respondents. The

Designated Officer for settling the claims and objections will be appointed by the State/UT governments. The Appellate Officers to settle appeals that come up over the decisions of the designated officers will also be appointed by the State/UT authorities. The timelines for the various stages and procedures of the entire claims and objections processes have been clearly earmarked and communicated to the States/UTs.

COTS modules receive all the claims and do the correction as per the decision made by the competent authority and finally upload the corrected data on central server in F files (database files). In case of rejection of claim, the respondent may appeal to appellate officer for further processing.

Final list generation: Once all the claims (with F files) were uploaded on central server, then came the stage of publishing the final data. Here NIC generates Z file (database files) which were used by CPSU team to generate PDF files of final list. Latest final list module used was 2.0.2.

Activity Based Software Support

Besides the process related software support at TPC and server level, there were few activities that needed software solutions. These are explained below:

(i) *Merging of SECC-NPR Data*: Initially it was decided to generate the draft list using cropped JPG images, but subsequently it was found that most of the images were not readable due to variant of handwriting and poor quality. Hence it was finalised to extract the corresponding digitised data from NPR database by merging the records on mandated fields like date of birth, gender and other unique fields for matching. The other challenges faced during merging of SECC and NPR databases were subsequently handled. As a result, merged data increased to 85 per cent of all households. The details on the merging of NPR and SECC database are explained in chapter 10.

(ii) Still the draft list contains 15 per cent of data with cropped images. Subsequently this 15 per cent images are decided to be digitised manually to make the full data set digitised and thus facilitating generation of all types of reports.

- *Digitisation of new member*: During survey, new members were found which were not available in NPR schedule and these new members were recorded in English digitised format. Later requirement was felt to convert this data in local language and hence this activity was also carried out by CPSUs.

Training and Capacity Building

Like any other operation, the training and capacity building of manpower that works with TPCs, servers are a prerequisite. The major task was to train the huge number of operator/database administrators for various types of software used in SECC. 6.4 lakh TPCs were used for the execution of SECC which in turns needed the same quantity of DEO to handle the TPC in field; hence training of these HHD was indeed very much required along with the hands on for the software developed for data entry. Training a small number of users is comparatively easy, but when that number goes up to 6.4 lakh then it will give some sort of stress on your head. Major challenge was to train the DEO for different type of modules required in SECC for short duration. Specifications of TPCs used in SECC are (details are in chapter 8):

Processor speed: 456 MHz.
Memory (RAM): 256 MB DDR2.
Storage (Internal): 128MB NAND and 2 GB Micro SD.
Storage (External): SD card support up to 32 GB.
Peripherals: USB ports (2 nos.).
Audio out: 3.5 mm jack for external speaker.
Audio in: 3.5 mm jack for integrated MIC.
Display: 7” LCD colour display with resistive touch.
Resolution: 800x480.
Input device: touch screen through stylus, keyboard.
Connectivity and networking: Ethernet 10/100 Mbps (RJ45).
Battery: 4000 mAH – 4 Hours.
Power adaptor: 5V, 1.5A.

Operating system: Android version: 2.2 - Froyo.

Operating temperature: 0 ° C TO +50 ° C.

Storage temperature: 0 ° C TO +70° C.

Dimension: 207 X 120 X 20 mm.

Weight in grams: 480 (Stand Alone), 820 (With Key Board).

Along with 6.4 lakhs TPC, 6.4 Lakhs of additional battery packs and 60,000 solar panel were used in SECC. Solar panel was introduced for use in difficult terrains where power not be available for charging the TPCs and additional battery packs. Capacity building in operating the software used in TPCs and local servers was achieved through training, but also for the basic use and maintenance of the accessories.

Conclusions

It was for the first time, a large scale survey for collecting socioeconomic data of each household was planned to be conducted using lakhs of TPCs, and thousands of local servers spread over the length and breadth of a country like India and in the end aggregating all these data in a central server was challenging. Though the rural and urban schedules to be canvassed did not undergo any change in the curse of survey, many predetermined processes were introduced to accommodate ground realities and demands of stakeholders. New processes like 'verifications and corrections', tracking of claims and objectors were introduced. Few critical modules were introduced such as for data viewing, draft and final list generation etc. Client software and server software were changed to accommodate demands of states on location Id and language of displaying the draft-list. Due to long time periods of survey involved, the charge centres located at sub-district level were closed in most of the locations and new facilities were arranged at district headquarters to facilitate uploading the survey outcomes. All these necessitated updating of the software continuously over the life of the project and solutions found to address all exigencies.

Because the SECC is a new venture and the hurdles that came up during conduct were not predictable, there was a need to go cautiously and adopt incremental corrective steps. Field realities were not fully grasped, but the project was rolled out. Hardly any pilots executed to test the veracity of the technologies and its consequences were undertaken. As a result, the project ran into long delays and cost overrun. However, all the core stakeholders worked with single minded objective of conducting the SECC to capture quality data of a billion and more individuals of the country. In the end it has been proved that such a big, complex and diverse project can be executed successfully.

10

PARTHA PRATIM BHATTACHARJEE

National Population Register

SECC Data Merging Procedure and Consequences

It was decided that Socio-economic and Caste Census (SECC) draft list will be published in text format and also in bilingual format implying that in both English as well as local language. Printing of Draft list will be done in local language and it will be displayed in Panchayat Office and other prominent places. For that, proper merging of NPR-SECC database is essential so that individual identification information will come in the database. It is here proposed to discuss the merging process in brief. Although it is a technical process, some basic idea would help the readers to appreciate the SECC work.

One would obviously like to ask why the merging process is required. It is because; the column 2 to column 9 information of SECC questionnaires will have to come from earlier collected NPR data. This mainly consists of the identification information of individuals, like—name, relationship to household head, sex (gender), date of birth, marital status, father's and mother's name, occupation etc. Although two fields of information are being collected again in SECC process, namely, sex (gender) and year of birth. It is clear that without these NPR data particular columns, the SECC database table will remain blank and it would not be possible to identify an individual without these NPR data.

The next question one would like to ask is how NPR and SECC data is merged together. In this context, merging refers to some portion of NPR database being copied/inserted/replicated to SECC database table in specific columns. For understanding of NPR-SECC merging procedure, one should have good understanding of NPR database.

Initially Office of the Registrar General & Census Commissioner, India (ORGI) made 'house listing' for the entire country for individual household to continue the Census work. According to that the entire country has been divided into 'HLBs (house listing blocks)'. These HLBs are the smallest geographical area of census operation specified by ORGI. According to that ORGI made a list of HLB for the entire country, which is known as the 'House Listing Block Master' or the house listing database. From year 2008 to 2010, the location database was made in the '*Janaganana*' i.e. National Population Register (NPR). In this method ORGI used papers for collecting information of each individual including their own household parameters. These pieces of paper have been named as HCS (housing census schedule) or NPR schedule. These NPR sheet/schedule were carried to each household by Census personnel. They had filled up these sheets/forms individually by hand. Now this hand written information is required to be digitised. Otherwise it will not be understood by computer.

Matching SECC-Enumeration Block (EB) and Population Enumeration (PE)-EB

After that ORGI made 'PE-2011', which was completed on March 2011 in the entire country. In PE phase ORGI divided entire country into 'EBs (enumeration blocks), which is the smallest area of Census specified by ORGI. It may be inhabited or uninhabited. According to that ORGI made a list of EB for the entire country, which is 'Enumeration Block Master' (EBM) or EBM Database. It is similar to the House Listing Block Master except that some HLBs are sub-divided into sub-blocks in PE phase. HLB are 4 digit codes and EB sub-blocks are 5 digit codes where last digit is sub-block code, which is 0 if it is not sub-divided and 1 to 9 as per sub-division. In PE phase, households information was collected and AHL (abridge household list) was prepared. It is nothing but EB wise list of households along with household head name and member count of that particular household, which are also handwritten. In a particular EB, 100 households may be taken as thumb rule. In AHL there are different sections for population—Section 2 households are the list of households which were present in NPR stage, Section 3 households are the list of households which are new households added in PE phase, Section 4 households are the houseless households. It is clear that the AHL Section 3 and 4 households haves no NPR sheets. In that particular household, only household head names had been entered in AHL.

In June 2011, SECC field enumeration was launched in Hazemara Block of Tripura State. Some basic information for SECC was required and hence, it was decided to use the AHL for individual household information and PE-EB Master for location information. But this EB Master has some problems and for progress of SECC enumeration this EB Master along with EB count has been changed rapidly as per request received from the field and through the State government. In September 2013, ORGI announced correct PE-EB Master, after one and half years of actual PE, whereas NPR started in 2008 (i.e. almost 5 years back). By this time the SECC EB Master has already gained its own shape which has completely deviated from initial EB Master and also from ORGI suggested new PE-EB Master. Now it becomes a major bottleneck for SECC and it starts delaying for matching of these two EB Master and it is necessary also for matching NPR database. SECC EB Master correction and matching with new PE-EB Masters becomes a tedious job for ORGI and also for NIC, because this SECC EB Master is kept only in NIC Server in soft form and this becomes another challenge. After long efforts, a significant percentage of PE-EB Master and SECC EB Master has been matched as well as mapped.

Matching of NPR database and SECC database is done in two pass matching method. First of all it has to match with NPR HL Master to PE-EB Master then PE-EB Master to SECC EB Master, which is also a tedious procedure and it becomes a big challenge ultimately to match the NPR House Listing Master database to SECC EB Master database. Here we may assume that ORGI is able to match with NPR HL Master to PE EB Master completely. In case the SECC EBM and PE-EB Master is not matched properly, then some EBs are extra in PE list and some EB are extra in SECC list. In NIC monitoring site, these extra and unmatched EBs were displayed as 'unenumerated EB' in its' dashboard, but it may be actually in most of the cases these EBs are unmatched EB in PE to SECC Master. After matching of EB level or

HLB level, then 2nd stage/pass—after that it has to match household members inside the matched HLB of NPR with EB sub-block of SECC.

Matching Household Members

Matching of household members inside the matched EB database is done through the following procedure. First it will match with database key of individual household member. In this process it has to be linked with the 27 digit NPR TIN (temporary identification number) of NPR database of matched HLB to 27 digit NPR TIN and 29 digit AHL TIN of SECC database of corresponding EB(s), which is/are already matched with that HLB. After that in 2nd pass, it will match 'YoB (year of birth)' and 'gender' (sex) of household members of matched SECC EB (it may be required for matching with member serial number of particular matched household). The two parameters/information mentioned above are common in both NPR and SECC, which are also in column 2 to 9 of SECC questionnaires. This matching is only possible in those household members which are in NPR (HCS) that means members of AHL Section 2 households (PE phase) to the household members who are present/available (member status 'AV') in SECC Enumeration process (AHL Section 2 household members). These members have proper NPR entry. No other household members can possibly be matched with NPR and SECC. This is the direct matching method.

When first method fails due to mismatch of location code of PE-EB Master and SECC EB Master, in that case FOR-READ method is adopted by ORGI. If any location code of 20 digit unique SECC EB Sub Block code is changed/not matched with corresponding PE Master EB Sub Block code, then it will not match the database. In that case FOR one EB of SECC EB Master is linked with READ of corresponding another one EB of PE-EB Master. Here one or more fields of location code mismatch can be linked. After the matching of EB, it follows above method for member matching by tagging/linking the corresponding location codes in TINs inside the EB through FOR-READ database in 2nd pass.

When above two methods, i.e. 1st and 2nd method, fail then the 3rd method i.e. 'Concordance' method has been adopted. The Concordance is nothing but FOR 'X' EB of SECC EB Master can be READ as 'Y' EB of PE EB Master, where 'X' and 'Y' both may be a single EB or a Group of EBs. In this case one or more places location code mismatch can be found. In this case 'Y' generally a single EB and 'X' may be a group of EBs. It will be very difficult if X and Y both are denoting more than 1 EB. General cases of Concordance 'Y' include cases where 1 and 'X' is more than 1. After the matching of EB it follows 1st method for member matching by tagging/linking the corresponding location codes in TINs inside the EB through FOR-READ database in 2nd pass.

Summarising the above method—first it has to match with SECC EB Master to NPR HLB Master then inside the EB Sub Block it has to match/link with individual 'AV' members of SECC to NPR individuals by using NPR TIN and AHL TIN along with 'Sex' and 'YoB' information. After this one to one match of some portion of NPR database (table) is copied/inserted/replicated to a particular portion of SECC database (column 2 to 9 information of SECC questionnaires). After that SECC database is assumed to be complete. This complete method is called as merging between NPR and SECC, or simple Merging NPR-SECC. In this method other databases like AADHAAR database, MGNREGS database or PMAY-G databases etc can also be merged with SECC database. Hence it will form a unique database for the entire country and DBT (direct benefit transfer) can be made possible for the beneficiaries.

Merging of NPR Text Data to SECC Database

It is clear that to finalise SECC database, some portion of NPR database has to be merged with SECC database. It is known that all NPR sheets (HCSs) are handwritten from field. Actually these NPR sheets are scanned in both sides (side A and B) by using a special scanner. After that, these 25 NPR sheets with both sides were put together in a single file in *.tiff file format. These *.tiff files or image files are then sent to different vendors appointed by ORGI for text conversion or digitisation of NPR sheet. Since all the scanned sheets or images are handwritten information in local language, they are not easy to digitise. Initially ORGI tried to use ICR (intelligent character recognition) software, but success rate has been found to be very low less than 50 per cent. Then they have used manual conversion method. In early 2011 when SECC started at that time only this digitisation was in starting stage, although the scanning of NPR images was all most ready. These vendors of ORGI converted this handwritten information to text data in two languages—one is local language for that particular State and another one is English. Finally it becomes the NPR database with these text converted data. Now the accuracy of these data should be cleared when an individual tries to convert hand written local language information into text data by typing in computer, sitting in different state. This is due to this fact that the quality of digitised data of some state is very poor.

These NPR scanned image files in *.tiff format has been used for SECC at the very beginning in local charge data centre. These *.tiff image files with 25 x 2 images have been cropped and used for SECC. From there Side A of NPR image has been segregated and structure reformatted and converted to 25 *.jpg image files, by using Cataloging software, which is provided by BEL. Actually *.jpg image format can be recognised by all type of hardware devices like camera, toy, TPC etc. In a particular NPR sheet, maximum eight household members can be placed, if any household have more than eight members then next sheet will be used, but in a single sheet more than one household is not entered that means rest are kept blank. Similarly in *.tiff file maximum 25 NPR sheets (both Side A & B) are placed but in a single file more than one scanned EB data will not be placed. If an EB having more than 25 NPR sheets is required then next *.tiff file will be generated and last one may be less than 25 sheets, as per population of an EB. This process of converting multiple image *.tiff file to single image compressed *.jpg file (Side A) is called as Cataloging method. These *.jpg files is installed in TPCs for SECC enumeration. Without these cataloged images of households, enumeration through TPCs is not possible. Actually the household serial number and some portion of household head name along with location information are attached to image filenames as a tag (refer to Figure A-10.1, of Annexure A-10.1). When these images are installed in TPCs, SECC Client software collect location tags from image files. Hence it is not possible to start SECC enumeration work by TPCs without images. In case of AHL Section 3 and 4 households and the households of Coastal area where no NPR images are available, in such cases dummy image files are used. In this cases it should be noted that above mentioned households including any new household found in SECC enumeration process or any new household members of existing household (AHL Section 2 household) are found, then that particular new members' status will be 'new' ('NW') and entire SECC questionnaires including column 2 to column 9 information also have to enter freshly in SECC enumeration or supervision phase. It was decided that after cataloguing all images, it has to be uploaded in NIC central server because in SECC database, image file name tags are also included in proper column. Thus it is clear that 100 per cent image file has to be prepared and uploading to central server is also essential just after cataloguing, otherwise some vital identification of household member will not be available. But practically in most of the places, image files have not been uploaded after cataloguing. In case where the text data has not come and images are not available, no member identification information is available and hence, Draft list is not possible to be generated for those particular households. It may be noted that it is not possible to get 100 per cent text converted data from ORGI for various reasons. This is because 100 per cent SECC AV members will not merge with NPR text data. Then only one source to re-trip individual member information is the NPR images. In draft list generating software it is tagged with images and the name portion from image is cropped and pasted to respective place of the draft list. It can be assumed that 80 per cent household members' text/digitised data will come from NPR database to SECC database and for remaining 20 per cent household members' text/digitised data has to be generate by MoRD itself as a post-facto digitisation work, which can only be done through NPR images.

Consequences

The advantage of relying on PE-2011 data collected for NPR as census frame (like sampling frame) for the SECC operations can't be denied. As mentioned in chapter 1, the advantage of using this process is much more prominent in urban areas than in rural areas. However, when the base data is in image format rather in digital format, many technical issues arise. Two such issue are: (i) matching PE-EB as collected under PE/NPR by ORGI with SECC-EB along with households and its members, and (ii) converting images of HCS or NPR schedules to digital formats. SECC being a paperless exercise, these two issues posed major challenges to the SECC project team. Though solutions to such problems could be found, implementations of these solutions resulted in inordinate delay in the finalising of SECC.

Annexure A-10.1

Nomenclature of Some Special Denoters

In the SECC database, several denoters are being used such as the database file name itself along with TINs is some example of it. Some of the most important denoters are given below so that readers can understand entire systems properly:

Figure A-10.1

Nomenclature of Image (.jpg) File Format*

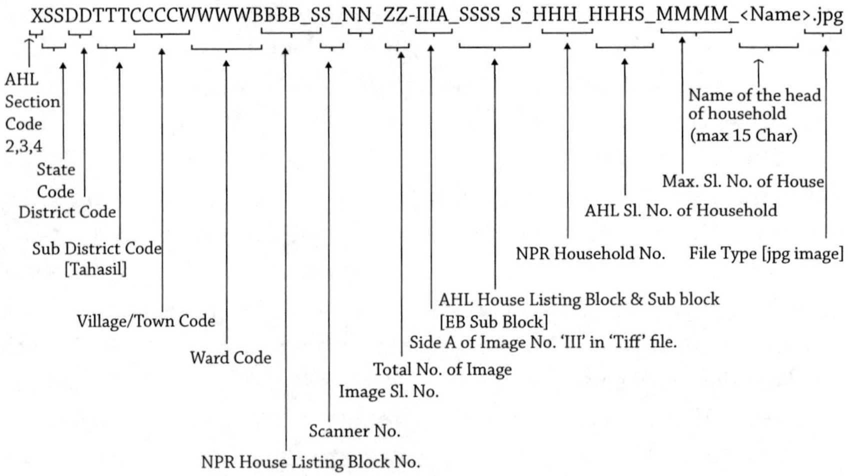

Example (UP, Ghaziabad):-
20909002710600770445_08_01_05-003A_0445_0_003_0003_0025_SANJU.jpg

Figure A-10.2

Nomenclature of Database (.db) File Format*

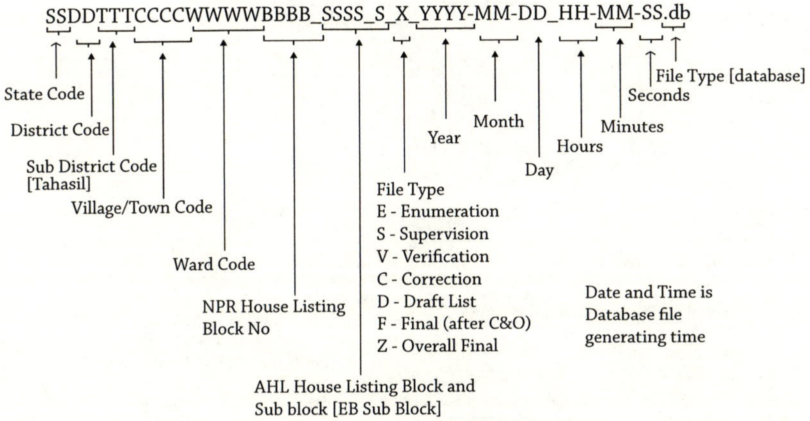

Example :-
0909002710600770445_0445_0_E_2012-12-13_13-23-51.db

Figure A-10.3

Nomenclature NPR TIN Format

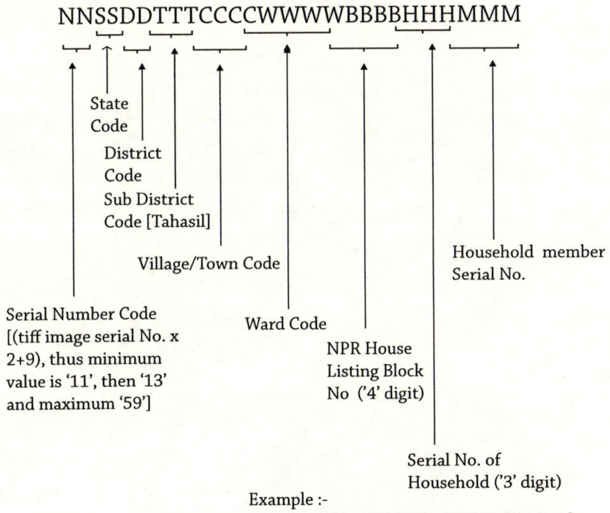

Example :-
530909002710600774440210O2 [27 digit]

Figure A-10.4

Nomenclature AHL TIN Format

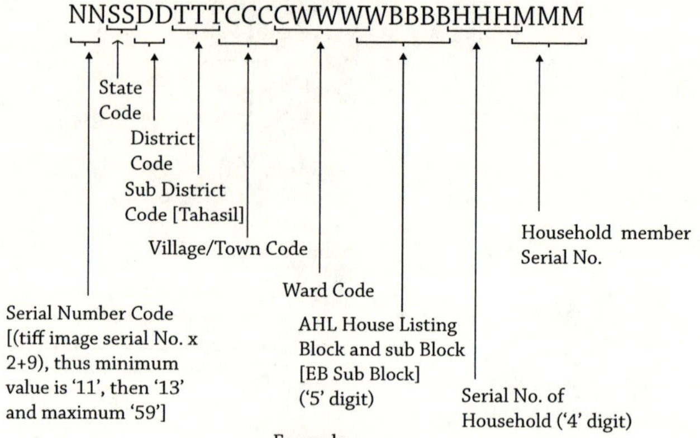

Example :-
53090900271060077044400021002 [29 digit]

Annexure A-10.2

Comparison with NPR and SECC Data

House listing and Housing Census Schedule, Census of India 2011 vs. SECC Household Schedule [SECC Side B questionnaires]	
NPR schedules as per Census of India 2011 [refer to Annexure 6.5 & 6.6, NPR Housing Census Schedules]	SECC schedules as per SECC 2011 questionnaires [refer to Annexure 6.3 & 6.4, SECC Rural & Urban questionnaires]
Predominant material of the floor, wall and roof of the Census house [Column 4 to 6]: (Give code number from the respective lists below).	Housing /Dwelling [Side B of Section - 1, Column 1 to 2, of SECC Rural questionnaires]: Housing /Dwelling [Side B of Section - 1, Column 1 to 2, of SECC Urban questionnaires]:
Floor [Column 4]: Mud (Code 1) Wood/bamboo (Code 2) Burnt brick (Code 3) Stone (Code 4) Cement (Code 5) Mosaic/floor tiles (Code 6) Any other (Code 7).	No schedule question has been asked in SECC.

Wall [Column 5]: Grass/thatch/bamboo etc. (Code 1) Plastic/polythene (Code 2) Mud/unburnt brick (Code 3) Wood (Code 4) Stone not packed with mortar (Code 5) Stone packed with mortar (Code 6) G.I./metal/asbestos sheets (Code 7) Burnt brick (Code 8) Concrete (Code 9) Any other (Code 0).	**Predominant material of wall of the dwelling room (s) (Give code) [Column 1]:** Grass/thatch/bamboo etc. (Code 1) Plastic/polythene (Code 2) Mud/unhurt brick (Code 3) Wood (Code 4) Stone not packed with mortar (Code 5) Stone packed with mortar (Code 6) G.I./metal/asbestos sheets (Code 7) Burnt brick (Code 8) Concrete (Code 9) Any other (Code 0).
Roof [Column 6]: Grass/thatch/bamboo/wood/mud etc. (Code 1) Plastic/polythene (Code 2) Handmade tiles (Code 3) Machine made tile (Code 4) Burnt brick (Code 5) Stone (Code 6) Slate (Code 7) G.I./metal/asbestos sheets (Code 8) Concrete (Code 9) Any other (Code 0).	**Predominant material of roof of the dwelling room (s) (Give code) [Column 2]:** Grass/thatch/bamboo/wood/mud etc. (Code 1) Plastic/polythene (Code 2) Handmade tiles (Code 3) Machine made tile (Code 4) Burnt brick (Code 5) Stone (Code 6) Slate (Code 7) G.I./metal/asbestos sheets (Code 8) Concrete (Code 9) Any other (Code 0).
Ascertain use of Census house [Column 7]: (Write the actual use and then choose the appropriate code from the list below and record the same in the box at the left hand side of the column) Residence (Code 1) Residence-cum-other use (Code 2) Shop/Office (Code 3) School/College etc. (Code 4) Hotel/lodge/guest house etc. (Code 5) Hospital/dispensary etc. (Code 6) Factory/workshop/workshed etc. (Code 7) Place of worship (Code 8) Other non-residential use (Code 9) Vacant (Code 0).	No schedule question has been asked in SECC.
If Code '1' or '2' in Column 7, condition of this Census house [Column 8]: Good (Code 1) Livable (Code 2) Dilapidated (Code 3).	No schedule question has been asked in SECC.
Total Number of persons normally residing in this household: Persons [Column 10]: Male [Column 11]: Female [Column 12]:	This information can be computed through the member status collected through TPC at the time of SECC Persons = Member Status -> AV + NW [Note MI, MO & DE are not the same household member] After that through 'Sex' code 'Male' and 'Female' can be computed.

If SC* or ST* or Other [Column 15]: *for col.15 Scheduled Caste can be only from Hindus, Sikhs and Buddhists and not from any other religion. Scheduled Tribe can be from any religion. SC (Code 1) ST (Code 2) Other (Code 3) **Applicable for Household head only.**	**Caste/Tribe Status:** [Side A of Section - 1, Block D, Column 13 & 14, of SECC Rural questionnaires]: [Side A of Section - 1, Block D, Column 16 & 17, of SECC Urban questionnaires]: Give Code: Scheduled Caste (SC) (Code 1) Scheduled Tribe (ST) (Code 2) Other (Code 3) No Caste/Tribe (Code 4) **(SC can be only among the Hindus, Sikhs and Buddhists. ST can be from any religion).** If code 1, 2 or 3 in Column 13 in Rural and Column 16 in Urban, Write Name of Caste/Tribe in text box in Column 14 in Rural and Column 17 in Urban If code 4 in Col. 16, put 'X' in Column 14 in Rural and Column 17 in Urban, #1.
Ownership status of this house [Column 16]: Owned (Code 1) Rented (Code 2) Any other (Code 3).	**Ownership status of this house (Give code):** [Side B of Section - 1, Column 3, of SECC Rural questionnaires]: | [Side B of Section - 1, Column 3 SECC Urban questionnaires]: Owned (Code 1) Rented (Code 2) Any other (Code 3).
Number of dwelling rooms # exclusively in possession of this household (Record 0, 1, 2, 3 ...) [Column 17]: # for col.17 Do not include kitchen, bathroom, latrine, store room, passageway and verandah. Include living room, bedroom, dining room, drawing room, study room, servant's room and other similar rooms.	**Number of dwelling rooms exclusively in possession of this household (Record 1,2,3...):** [Side B of Section - 1, Column 4, of SECC Rural questionnaires]: [Side B of Section - 1, Column 4 SECC Urban questionnaires]
No schedule question has been asked in Census.	**Is any household member:** from Primitive Tribal Group; a legally released bonded labour; a manual scavenger. [Side B of Section - 2, Column 5, 6 & 7 of SECC Rural questionnaires]: Yes (Code 1) , No (Code 2), #2.
Amenities available to the household [Column 19 to 27]:	Amenities [Side B of Section - 2, Column 5 to 9, of SECC Urban questionnaires]
Main Source of Drinking Water [Column 19]: Tap water from treated source (Code 1) Tap water from un-treated source (Code 2) Covered well (Code 3) Uncovered well (Code 4) Hand pump (Code 5) Tube well /Borehole (Code 6) Spring (Code 7) River/canal (Code 8) Tank /pond /lake (Code 9) Other sources (Code 0).	No schedule question has been asked in SECC.

Availability of drinking water source [Column 20]: Within the premises (Code 1) Near the premises (Code 2) Away (Code 3).	**Availability of drinking water source [Section 2, Column 5 of SECC Urban questionnaires]:** Within the premises (Code 1) Near the premises (Code 2) Away (Code 3), #3.
Main source of lighting [Column 21]: Electricity (Code 1) Kerosene (Code 2) Solar (Code 3) Other oil (Code 4) Any other (Code 5) No lighting (Code 6).	**Main source of lighting [Section 2, Column 6 of SECC Urban questionnaires]:** Electricity (Code 1) Kerosene (Code 2) Solar (Code 3) Other oil (Code 4) Any other (Code 5) No lighting (Code 6), #3.
Latrine within the premises [Column 22]: Yes (Code 1) No (Code 2). If 'Yes' (Code 1) in Column 22, give Code from '1' to '8'; If 'No' (Code 2) in Column 22, give Code '9' or 0 from the list below [Column 23]: **Flush/pour flush latrine connected to:** Piped sewer system (Code 1) Septic tank (Code 2) Other System (Code 3). **Pit latrine:** With slab /Ventilated improved pit (Code 4) Without slab/open pit (Code 5). Night soil disposed into open drain (Code 6) **Service latrine:** Night soil removed by human (Code 7) Night soil removed by animals (Code 8). **No latrine within the premises :** Public latrine (Code 9) Open defecation (Code 0).	**Water-seal latrine exclusively for the household [Section 2, Column 7 of SECC Urban questionnaires]:** Yes (Code 1) No (Code 2), #3.
Waste water outlet connected to [Column 24]: Closed drainage (Code 1) Open drainage (Code 2) No drainage (Code 3).	**Waste water outlet connected to** [Section 2, Column 8 of SECC Urban questionnaires]: Closed drainage (Code 1) Open drainage (Code 2) No drainage (Code 3), #3.
Bathing facility available within the premises [Column 25]: Yes : Bathroom (Code 1) Enclosure without roof (Code 2) No (Code 3).	No schedule question has been asked in SECC.
Availability of Kitchen [Column 26]: Cooking inside house: Has kitchen (Code 1) Does not have kitchen (Code 2) Cooking outside house: Has kitchen (Code 3) Does not have kitchen (Code 4) No cooking (Code 5).	**Separate room used as kitchen exclusively for the household [Section 2, Column 9 of SECC Urban questionnaires]:** Yes (Code 1) No (Code 2), #3.

Fuel used for cooking [Column 27]: Firewood (Code 1) Crop residue (Code 2) Cow dung cake (Code 3) Coal /lignite /charcoal (Code 4) Kerosene (Code 5) LPG /PNG (Code 6) Electricity (Code 7) Biogas (Code 8) Any other (Code 9) No cooking (Code 0).	No schedule question has been asked in SECC.
Assets possessed by the household [Column 28 to 34]:	Assets [Side B of Section - 4, Column 14 to 16, of SECC Rural questionnaires]: Assets [Side B of Section - 3, Column 10 to 15, of SECC Urban questionnaires]:
Radio /transistor are available [Column 28]: Yes (Code 1) No (Code 2).	No schedule question has been asked in SECC.
Television are available [Column 29]: Yes (Code 1) No (Code 2).	No schedule question has been asked in SECC.
Computer/laptop are available [Column 30]: Yes: With internet (Code 1) Without internet (Code 2) No (Code 3).	**Computer/laptop are available [Side B of Section - 3, Column 12, of SECC Urban questionnaires]:** Yes: With Internet (Code 1) Without Internet (Code 2) No (Code 3), #3.
Telephone/mobile phone are available [Column 31]: Yes: Land line only (Code 1) Mobile only (Code 2) Both (Code 3) No (Code 4).	**Telephone/mobile phone are available [Side B of Section - 4, Column 15, of SECC Rural questionnaires and Side B of Section - 3, Column 11, of SECC Urban questionnaires]:** Yes: Land line only (Code 1) Mobile only (Code 2) Both (Code 3) No (Code 4). No schedule question has been asked in SECC.
Bicycle are available [Column 32]: Yes (Code 1) No (Code 2) **Scooter /Motor Cycle /Moped are available [Column 33]:** Yes (Code 1) No (Code 2). **Car /Jeep /Van are available [Column 34]:** Yes (Code 1) No (Code 2)	**Motorised two/three/four wheelers or motorised fishing boat requiring registration are available [Side B of Section-4, Column 16, of SECC Rural questionnaires]:** No vehicle (Code 1) Two wheelers only (Code 2) Three wheelers only (Code 3) Four wheelers only (Code 4) Fishing boat (Code 5). Motorised wheelers are available [Side B of Section - 3, Column 13, of SECC Urban questionnaires]: Two/three wheeler (Code 1) Four wheeler (Code 2) No (Code 3).

No schedule question has been asked in Census.	Refrigerator are available [Side B of Section - 4, Column 14, of SECC Rural questionnaires and Side B of Section - 3, Column 10, of SECC Urban questionnaires]: Yes (Code 1) No (Code 2).
No schedule question has been asked in Census.	A. C. are available [Side B of Section - 3, Column 14, of SECC Urban questionnaires]: Yes (Code 1) No (Code 2).
No schedule question has been asked in Census.	Washing machine are available [Side B of Section - 3, Column 15, of SECC Urban questionnaires]: Yes (Code 1) No (Code 2). Other Assets owned [Side B of Section - 5B, Column 21 to 23, of SECC Rural questionnaires]:
No schedule question has been asked in Census.	Mechanised Three/Four Wheeler Agricultural equipment are available [Side B of Section - 5B, Column 21, of SECC Rural questionnaires] Yes (Code 1) No (Code 2), #2.
No schedule question has been asked in Census.	Irrigation equipment (including diesel/kerosene/electric pump set, sprinkler/drip irrigation system, etc.) are available [Side B of Section - 5B, Column 22, of SECC Rural questionnaires]: Yes (Code 1) No (Code 2), #2.
Available Banking Services [Column 35]: Yes (Code 1) No (Code 2).	Kisan Credit Card with credit limit of Rs 50,000 or above are available [Side B of Section - 5B, Column 23, of SECC Rural questionnaires] Yes (Code 1) No (Code 2), #2. Land owned [Side B of Section - 5A, Column 17 to 20, of SECC Rural questionnaires]
No schedule question has been asked in Census.	Own any land (excluding homestead)? If landless, skip to Section-5B [Side B of Section - 5A, Column 17, of SECC Rural questionnaires]: Yes (Code 1) No (Code 2), #2.
No schedule question has been asked in Census.	If 'Yes' Total un-irrigated land (in acres) [Side B of Section - 5A, Column 18, of SECC Rural questionnaires], #2.
No schedule question has been asked in Census.	If 'Yes' Total irrigated land (in acres) With assured irrigation for two crops [Side B of Section - 5A, Column 19, of SECC Rural questionnaires], , #2.

No schedule question has been asked in Census.	If 'Yes' Other irrigated land (in acres) [Side B of Section - 5A, Column 20, of SECC Rural questionnaires], #2. Employment and Income Characteristics [Side B of Section - 3, Column 8 to 13, of SECC Rural questionnaires]:
No schedule question has been asked in Census.	Does any household member: Have a salaried job? [Side B of Section - 3, Column 8, of SECC Rural questionnaires]: Yes (Code 1) No (Code 2), #2.
No schedule question has been asked in Census.	If 'Yes' in Col. 8, Salaried job is in (give code) [Side B of Section - 3, Column 9, of SECC Rural questionnaires]: Government (Code 1) Public sector (Code 2) Private sector (Code 3), #2.
No schedule question has been asked in Census.	Does any household member: Pay income tax or professional tax [Side B of Section - 3, Column 10, of SECC Rural questionnaires]: Yes (Code 1) No (Code 2), #2.
No schedule question has been asked in Census.	Does any household member: Own/operate an enterprise which is registered with the Government [Side B of Section - 3, Column 11, of SECC Rural questionnaires]: Yes (Code 1) No (Code 2), #2.
No schedule question has been asked in Census.	Monthly income of highest earning household member (give code): [Side B of Section - 3, Column 12, of SECC Rural questionnaires] less than ₹5,000 (Code 1) between ₹5,000 and ₹10,000 (Code 2) ₹10,000 or more (Code 3), #2.
No schedule question has been asked in Census.	Main source of household income from (Give code) [Side B of Section - 3, Column 13, of SECC Rural questionnaires]: Cultivation (Code 1) Manual casual labour (Code 2) Part-time or full-time domestic service (Code 3) Foraging, rag picking (Code 4) Non-agricultural Own Account Enterprise (Code 5) Begging/Charity/Alms collection (Code 6) Others (Code 7), #2.

Household Schedule, National Population Register vs. SECC Individual Household member details [SECC Side A questionnaires]:	
Name of the person in full and resident status [Side A, Column Q.1]: 1(a) Write name of the person in full (Start with the head of the household), Name of the person [Side B, Column Q.1]: Copy from Q.1(a) of Side A in same order 1(b) Write resident status code in box Usual Resident (Actual) (Code 1) Usual Resident (Intention) (Code 2) Others (Code 3). **Name of the person as should appear in the NPR** [Side A, Column Q.2]: (maximum 30 boxes, abbreviate if required). **Name of the Head of the household** [Column 13]: (as per Census of India 2011).	Name of the person Start with the head of the household (maximum 30 characters, abbreviate if required) [Side A of Section - 1, Block D, Column 2, of SECC Rural questionnaires] [Side A of Section - 1, Block D, Column 2, of SECC Urban questionnaires] [populated from NPR]:
Relationship to head (record the relationship in full) [Side A, Column Q.3]:	Relationship to head (record the relationship in full) [Side A of Section - 1, Block D, Column 3, of SECC Rural questionnaires]: [Side A of Section - 1, Block D, Column 3, of SECC Urban questionnaires]: [populated from NPR].
Sex [Side A, Column Q.4]: Male (Code 1) Female (Code 2). **Sex** [Column 14]: (as per Census of India 2011) (Applicable for Household head only) Male (Code 1) Female (Code 2).	Sex : [Side A of Section - 1, Block D, Column 4, of SECC Rural questionnaires] [Side A of Section - 1, Block D, Column 4, of SECC Urban questionnaires] Male (Code 1) Female (Code 2) [populated from NPR] In Tablet PC Male (Code 1), Female (Code 2), Transgender (Code 3), has been captured.
Date of birth (as per English Calendar), Day, Month, Year [Side A, Column Q.5]:	Date of birth (as per English Calendar): [Side A of Section - 1, Block D, Column 5, of SECC Rural questionnaires]: [Side A of Section - 1, Block D, Column 5, of SECC Urban questionnaires]: [populated from NPR]. In Tablet PC Year of Birth data has been captured
Marital status [Side A, Column Q.6]: Never married (Code 1) Currently married (Code 2) Widowed (Code 3) Separated (Code 4) Divorced (Code 5). **Number of married couple(s) living in this household** (Recorded 0,1,2,3.....) [Column 18]: (as per Census of India 2011)	Marital status [Side A of Section - 1, Block D, Column 6, of SECC Rural questionnaires]: [Side A of Section - 1, Block D, Column 6, of SECC Urban questionnaires]: [populated from NPR]. Never married (Code 1) Currently married (Code 2) Widowed (Code 3) Separated (Code 4) Divorced (Code 5).

Educational qualification [Side A, Column Q.7]:	Highest educational level completed [Side A of Section - 1, Block D, Column 10, of SECC Rural questionnaires]: [Side A of Section - 1, Block D, Column 10, of SECC Urban questionnaires]: Illiterate (Code 1) Literate but below primary (Code 2) Primary (Code 3) Middle (Code 4) Secondary (Code 5) Higher secondary (Code 6) Graduate or higher (Code 7) Other (Specify) (Code 8).
Occupation/Activity [Side A, Column Q.8]: Describe the actual work.	Occupation/Activity: Describe the actual work. [Side A of Section - 1, Block D, Column 9, of SECC Rural questionnaires]: [Side A of Section - 1, Block D, Column 9, of SECC Urban questionnaires]: [populated from NPR].
Name(s) of father, mother and spouse in full [Side A, Column Q.9]: if father, mother and spouse are not enumerated in this household or are not alive write Father's name against 'F' Mother's name against 'M' Spouse's name against 'S' OR If they are enumerated in this household, write serial number of the father, mother and spouse in the boxes as recorded in Column 1.	**Name of father**: [Side A of Section - 1, Block D, Column 7, of SECC Rural questionnaires]: [Side A of Section - 1, Block D, Column 7, of SECC Urban questionnaires]: [populated from NPR] **Name of mother**: [Side A of Section - 1, Block D, Column 8, of SECC Rural questionnaires]: [Side A of Section - 1, Block D, Column 8, of SECC Urban questionnaires]: [populated from NPR].
Place of Birth [Side B, Column Q.10]: if within India, write the present name of the Village/town, District and state. if outside India, write the present name of the Country and put '-' against Village/town and District.	No schedule question has been asked in SECC.
Nationality as declared [Side B, Column Q.11]: Q.11 *NOTE: Nationality recorded in Q.11 is as declared by the respondent. This does not confer any right to Indian Citizenship. Indian (Code 1) Others write name of the Country.	No schedule question has been asked in SECC.

Present Address of usual residence [Side B, Column Q.12]: if the person resides or intends to stay for more that 6 months at this address (Write complete address including the (i) Building number and Name, House Number, (ii) Street name, (iii) Locality/Post Office, (iv) Village/Town (v) District, State), PIN (6 digit). **Building Number** [Column 2]: (as per Census of India 2011) (Applicable for Household schedule), #4.	No schedule question has been asked in SECC.
Duration of Stay at present address [Side B, Column Q.13]: (in complete years).	No schedule question has been asked in SECC.
Permanent Address of usual residence [Side B, Column Q.14]: if same as Q.12 write 'Same' otherwise (Write complete address including the (i) Building number and Name, House Number, (ii) Street name, (iii) Locality/Post Office, (iv) Village/Town (v) District, State), PIN (6 digit), #4.	No schedule question has been asked in SECC.
No schedule question has been asked in Census.	Main source of income/earnings from Work: [Side A of Section - 1, Block D, Column 11, of SECC Urban questionnaires]: Beggar/rag-picker (Code 1) Domestic worker (Code 2) Street vendor/cobbler/hawker /other service provider working on streets (Code 3) Construction worker/plumber/mason/labor/painter/welder/security guard/coolie and other head-load worker (Code 4) Sweeper/sanitation worker /mali (Code 5) Home-based worker/artisan/handicrafts worker / tailor (Code 6) Transport worker/driver/conductor/helper to drivers and conductors/cart puller/rickshaw puller (Code 7) Shop worker/assistant/peon in small establishment/helper/delivery assistant /attendant/waiter (Code 8) Electrician/mechanic/assembler/repair worker (Code 9) Washer-man/chowkidar (Code 10) Other work (Code 11) Non-work Non-work (Pension/Rent/Interest, etc.) (Code 12) No income from any source (Code 13), #3.

No schedule question has been asked in Census.	Are Wages earned (give code): [Side A of Section - 1, Block D, Column 12, of SECC Urban questionnaires]: daily (Code 1) weekly (Code 2) monthly (Code 3) ir-regularly (Code 4) not wage earner (Code 5), #3.
No schedule question has been asked in Census.	Disability: [Side A of Section - 1, Block D, Column 11, of SECC Rural questionnaires]: [Side A of Section - 1, Block D, Column 13, of SECC Urban questionnaires]: In Seeing (Code 1) In Hearing (Code 2) In Speech (Code 3) In Movement (Code 4) Mental Retardation (Code 5) Mental Illness (Code 6) Other disability (Code 7) Multiple disability (Code 8) Not disabled (Code 9).
No schedule question has been asked in Census.	Chronic illness: [Side A of Section - 1, Block D, Column 14, of SECC Urban questionnaires]: Cancer (Code 1) HIV/AIDS (Code 2) TB (Code 3) Leprosy (Code 4) Other illness (Code 5) No chronic illness (Code 6), #3.
No schedule question has been asked in Census.	Religion: (Write name of the religion in full) [Side A of Section - 1, Block D, Column 12, of SECC Rural questionnaires]: [Side A of Section - 1, Block D, Column 15, of SECC Urban questionnaires]: #7.
House-listing and Housing Census Schedule, Census of India 2011 and Household Schedule, National Population Register location information Vs. SECC Individual Household location information [SECC Side A questionnaires]	
State/UT name and Code (2 digit): **District** name and code (2 digit): **Tehsil/Taluk/P.S/Dev. Block/Circle/Mandal** (3 digit): **Town/Village** name and code (4 digit): **Ward Code No.** (Only for town) (4 digit): ["Location Parameters" of Side A of both NPR and Census of India 2011 Household Schedule]:	State/UT name and Code (2 digit): District name and code (2 digit): Tehsil/Taluk/P.S/Dev. Block/Circle/Mandal (3 digit): Town/Village name and code (4 digit): Ward code No. (Only for town) (4 digit): [Side A of Block A, of SECC Rural & Urban questionnaires]: #5.

House-listing Block Number (4 digit): [“Location Parameters” of Side A of both NPR and Census of India 2011 Household Schedule]	House-listing Block Number (4 digit): [Side A of Block B, of SECC Rural & Urban questionnaires]: Enumeration Block Number & Sub-Block No. (Item-6, Section-1) [4 + 1 digit) (To be copied from the Abridged House list (AHL)): [Side A of Block C, of SECC Rural & Urban questionnaires]:
Household Number (3 digit): (Give separate serial number to each household and write ‘999’ for every institutional household) [Column 9]: (as per Census of India 2011) [“Location Parameters” of Side A of both NPR and Census of India 2011 Household Schedule] **Census House Number** [Column 3]: (as per Census of India 2011)	Household Number (3 digit): [Side A of Block B, of SECC Rural & Urban questionnaires]: AHL Serial Number of Household (4 digit): (Column 8 of section 2 or 3 or column 6 of section 4, to be copied from AHL) [Side A of Block C, of SECC Rural & Urban questionnaires]:
Institutional can be completed as household serial number is ‘999’. AHL Section 4 households are the Houseless households.	Type of Household: [Side A of Section - 1, Block C, Column 12, of SECC Rural questionnaires] [Side A of Section - 1, Block C, Column 15, of SECC Urban questionnaires] Normal (Code 1) Institutional (Code 2) Houseless (Code 3) If institutional household, give details [Side A of Block A, of SECC Rural questionnaires] If ‘Houseless’, is the household living in shelter for homeless? [Side A of Block A, of SECC Urban questionnaires], #6.
Same is available in **Charge Register** for Census of India 2011 (PE):	Type of EB (Slum/Non-Slum) (For Statutory Towns): (To be obtained from col.17 of Supervisory Booklet of Charge Register for Population Enumeration 2011 (PE)) [Side A of Block B, of SECC Urban questionnaires]: #3. Only in SECC Urban Schedule

Details of the ‘Remarks’ :

#1. Caste Code (SC/ST) has been collected for entire household whereas in SECC individual Caste has been recorded along with Caste Name [only Alpha character and ‘@’, ‘,’, ‘’’, ‘ ’ are allowed].

#2. Only in SECC Rural Schedule.

#3. Only in SECC Urban Schedule.

#4. This information can be added in SECC database for value addition in future.

#5. Same as NPR & SECC

#6. Last one information collected Only in SECC Urban Schedule

#7. In SECC individual Religion has been recorded [only Alpha character and ‘@’, ‘,’, ‘’’, ‘ ’ are allowed].

11

VARINDRA SETH, PRAMOD SHARMA and N.K. SAHU

Database Management and MIS of SECC

Tango between Hardware and Software

All traditional civilisations are rich in mythology as well as history. Both in mythology and history evidence of records of events, characters and many more, such things are found systematically recorded. Benefits and working of such systematic records were recognised by these civilisations. In Indian tradition, the Hindu religious scriptures mentioned description of Chitragupta—a God assigned with the task of keeping complete records of actions of living beings on the earth is well recognised. Upon their death, Chitragupta has the task of deciding heaven or the hell for the humans and living beings depending on their actions on the earth. He assists the God of death, Yama, also known as Dharamraj or Yamraj in this regard. The records contains, according to some scriptures, 900,000 species living in the water, 2,000,000 of kinds of *sthāvara* (non-moving living entities such as trees and plants), 1,100,000 species of insects and reptiles, 1,000,000 species of birds, 3,000,000 varieties of quadrupeds and 400,000 human species. Interestingly, these records are dynamic as they are updated as per the actions of each one of these living beings. Such a database works as a decision support system to the God Yama, in arriving at who would go to heaven or to hell, with magnitude of rewards and punishments respectively. The modern day record keeping known as database management is aspiring to be like the kind of mythical database of living beings maintained by Chitragupta. In this chapter, the objective is to explain the nature and the database of households and individuals created for storing with analytics for the Socio-economic and Caste Census (SECC). This task was performed by National Informatics Centre (NIC), a Government of India (GoI) organisation.

Background

By the time it was planned to organise the SECC, the data collection under the decennial population census of India has been over. Thus what would be the size of the proposed SECC database was fairly known to the project team. Population Census 2011 and preparation of the National Population Register (NPR) pre-dated the SECC operations. While population census captured the demographic particulars of all population, the objective of the NPR was to create a comprehensive identity database of every usual resident in the country—a person who has resided in a local area for the past six months or more or a person who intends to reside in that area for the next six months or more. Since the Office of the Registrar General and Census Commissioner, India (ORGI) is one of the principal stakeholders in SECC project (caste data is collected as it could not be collected in population census operation), understanding the dimensions and complexities of the proposed database was facilitated by ORGI and Census Commission. Secondly, the database created by Ministry of Rural Development (MoRD), GoI in previous BPL (below poverty line) census including the BPL Census 2002 (Rural) was the light house for SECC team to design the database management of SECC. The additional data expected to populate the SECC database, besides the demographic and programme/scheme targets parameters, arises due to adoption of new and advanced method of data collections through use of TPCs (tablet personal computers) for data collection and online data transfer protocols. Such operations would generate lots of metadata and logs. Thus while designing the database management system, the above mentioned background information and technical exigencies determined the boundaries of the SECC database management.

Though most of the operations were known along with its' volume of data as well as metadata to be handled, a number of new processes and requirements were introduced subsequently as it was explained in previous chapters. Thus the nature and size of the database changed along with the project. As we know, for any database management the requirement is to balance between the hardware and the database that works as partners. Before we explain this aspect, we need to know the salient features of the database. Figure 11.1 would help us to understand this.

Figure 11.1

Salient Features of the Database

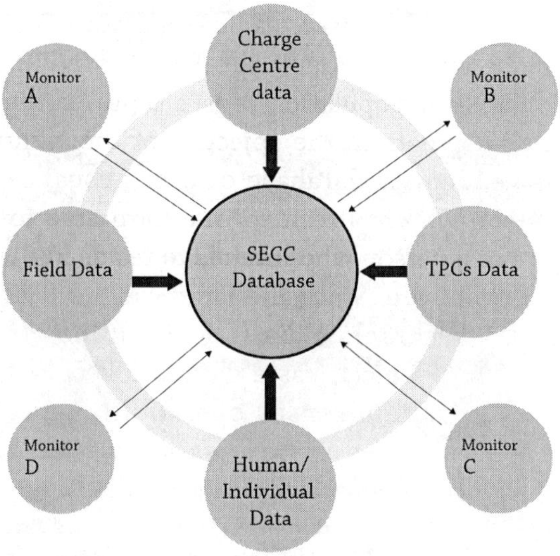

Core Database

From Figure 11.1, one can see that the core database consists of household and individual particulars collected from rural and urban areas. This is the database that contains the household and individual parameters based on which the government scheme beneficiaries would be identified. The details of the information captured, stored and put to analytics are given in Table 11.1.

Metadata

The electronically captured information/data also facilitate capturing of many aspects of the core database. Such data are valuable as it can be used to improve the quality of the core database that emanates from the field survey. As such data speaks about the events and processes, it is called metadata. These metadata may be digital in nature (identification numbers of households, individuals, etc.) or non-digital (instruction/training manuals etc). They may be real-time data or legacy data. Some of the important metadata captured while building the SECC database are explained in Table 11.1.

Table 11.1

Socioeconomic Information to be Collected Under SECC

Category	Rural	Urban
Identifications	State, District, Sub-district (Tehsil), Gram Panchayat, Village , House-list Block No, Household No,	State, District, Sub-district (Tehsil), Town, Ward code No, House-list Block No, Household No,
Demography	Type of HH (Normal/Institutional/ Houseless)	Type of HH (normal/institution-al/houseless)
	Name of person	Name of person
	Relationship to head of household	Relationship to head of household
	Gender	Gender
	Year of birth	Year of birth
	Father's/mother's name	Father's/mother's name
	Marital status	Marital status
	Occupation	Occupation
	SC/ST status	Sc/st status
	Disability	Disability
	Education	Education
	PTG	
	Legally released bonded labour	
	Manual scavenger	
		Type of block(slum/non-slum)
		Houseless- living in shelter
Housing / Dwelling	Wall type	Wall type
	Roof type	Roof type
	Ownership status	Ownership status
	Rooms (Numbers)	Rooms (Numbers)

Employment/ Income Characteristics	Household having employed member	
	Income/Professional tax paying HH	
	Household owning enterprise	
	Income of highest earner	Wage (Periodicity)
		Source of income
Assets	Refrigerator	Refrigerator
	Telephone/mobile phone	Telephone/mobile phone
	Motorized wheelers/boats	Motorised wheelers
	Agricultural equipment	
	Irrigation equipment	
	Kisan credit card	
		Air conditioner
		Washing machine
		Computer/laptop
Land	Land (Yes/No)	
	Total unirrigated land	
	Total irrigated land	
	Area under assured irrigation (two crops)	
	Other irrigated land	
Amenities		Water source
		Source of lighting
		Water-seal latrine
		Waste water outlet
		Kitchen (Separate)

TPC data: During the various processes of the census, the data related to the date and time of the operation was captured. These data are very vital and give valuable clues to the managers of the SECC. For example, EB (enumeration block) (a geographical unit of a village or an urban area) consisting of 100 to 150 households cannot be enumerated in a few hours or even couple of days. As per instructions/ guidelines/norms, an EB is to be covered in ten days using the TPCs by the enumerator and 10 DEO (data entry operator) duo. But any TPC reporting the duration of enumeration of a EB much less than 10 days arises doubts and prompts the managers to take appropriate actions against the identified enumerator and DEO. These metadata consists of: (i) Date and time of enumeration, (ii) Date and time of supervision, (iii) Remarks of the enumerator on the respondents answers, (iv) Remarks of the supervisor on whether he agrees with the view of the enumerator. The size of such metadata was very small, but the value provided by it was immense.

Data charge centre (DCC): As it was mentioned in chapter 9 "Software Development for SECC: The Challenges of Meeting Multiple Demands" that every development block or/and *tehsil* opened DCC for undertaking SECC related activities under the supervision of a charge officer. From here all TPCs were activated, cataloguing AHL were done and proper electrical charging facilities provided. During such process, both man and machines (local servers) were put to service with log-in protocols. In the local servers/PCs located in these DCC, the authenticated officers were facilitated to see the 'Data View Module' and make appropriate interventions with regard to the data collected by the enumerators. From here only the field supervisors get a set of sample households that were to be supervised with the help of TPCs. During this process, the metadata containing identification of the charge centres, log-ins and footprints of events were captured. And this metadata was of great value as it lights on the rigours put on by the manpower working in these DCCs under the control of the Charge Officers.

Human/individual metadata: Near about a million skilled manpower were engaged in the SECC works. The category of manpower, their qualifications and training imparted to them was explained in Chapter 1. Identification of these individuals was possible due to the metadata base that was created as electronic data capture through either actual registration details or simply capturing the logs. Juxtaposing such metadata with that as captured in the TPCs and DCCs helped in finding out pointers to negligence on the part of the individual and/or agencies. Timely warning to them was possible and actually done throughout all the stages of the SECC processes.

Metadata reports: Though a number of reports were generated and put to use for monitoring and improving the quality of SECC outcomes, few examples are given to highlight the utility of such an exercise:

Report on Time taken for enumeration and supervision stages: The time taken for completing the data entry in enumeration and supervision stages is identified based on latest uploaded data of that EB.

Report on Correction done in supervisor records having enumeration remarks: For the records in enumerator data which are having Enumerator' Remarks, to be checked if these records are covered in supervisory data and if covered having any correction done or not?

Report on uncovered households: There are a number of households which are not enumerated in either due to 'household not found' or due to reason like house locked/non-response etc. This report gives the number of households which were not enumerated during enumeration but got covered in supervision.

Report on updation done in supervision: No of records covered in both enumeration and supervision and data updated in supervision; and

Report on records flagged in enumeration: Out of records flagged for checking in enumeration, the number of records covered in supervision and data updating.

Monitoring: Monitoring of progress, quality and additional facilities created like comparisons between data collected in other projects including the Population Census 2011 has been carried-out with the use of metadata. For example, it was found that the coverage of scheduled tribes in state of Jharkhand differed very much as compared to Population Census. Deficient/surplus districts and the locations were identified and re-enumerations were undertaken in such locations.

SECC's Adjunct Databases

When two or more databases help each other to improve their quality and utility, their value goes up. Such possibilities were brightened with regard to SECC database as it was assisted by two more existing databases. They were, the BPL database of 2002 that contains socioeconomic parameters of all rural households owned by MoRD and the National Population Register & Census database of 2011 owned by the Registrar General and Census Commission of India.

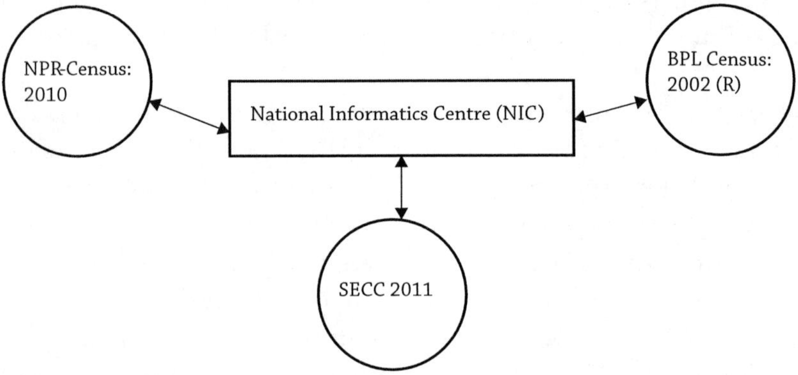

As the office of RGI & Census Commission was an active partner in the SECC project, assistance of all sorts, including household data etc., were forthcoming from them and thus, access to all three database was possible seamlessly.

Monitoring Reports

The SECC database facilitates monitoring of the process as well as the outcomes on a day-to-day basis. The monitoring web link *sec-census.nic.in* was created as a restricted facility for only the main stakeholders and their authorised persons. This facility provided: (i) summary status of progress of SECC at a particular point of time, (ii) census monitor, (iii) uploaded image status, (iv) verification and correction monitor, (v) claims monitor, (vi) miscellaneous reports, (vii) SECC results, etc. During the data collection, validation and use, this web based facility helped all stakeholders and its agents not only to monitor the progress in various field activities, but also the quality of data. 'Census Monitor' was the most useful tool that provided information on field operations like enumeration, supervision, verification etc. 'Claims Monitor' helped to monitor the progress in the most crucial element of SECC that aims at bringing in transparency and peoples' participation. Types of 'claims and objections' raised by affected parties, identification of such households/individuals/legal bodies and the redressal of such claims and objections were registered in this web-facility. Even one can monitor each individual case with the help of this facility. Besides these above mentioned facilities available in the monitoring site, a number of ready to use reports were generated and kept in this site to facilitate various operations of SECC. Briefs on these reports can be seen in the Annexure A-11.1.

Data Completeness and Quality Analysis

The data uploaded from the field was analysed for coverage, completeness and quality. For ensuring the coverage, the location directory of the ORGI was used. The ORGI location directory is a standard location directory where each village and EB can be identified with a unique code. For ensuring completeness and quality of the information apart from the local data viewing module, various data completeness/quality reports were facilitated to the authorised field authorities. The identified deficiencies were corrected by the field agencies through a process of re-uploading of corrected data. The process involved justification for re-uploading and an approval mechanism.

Hardware for SECC Database and MIS

For building a successful and useful database, the software and hardware should work hand-in-hand. Full utility of the database can only be maximised only when the complementary behaviour between them is identified and actions are taken accordingly. Anastasia Ailamaki, Professor of Computer and Communication Sciences at the Ecole Polytechnique Federale de Lausanne (EPFL) in Switzerland highlighted in the 41st International Conference on very large data bases in September 2015 that:

> In data management, system performance is defined as acceptable response time and throughput on critical-path operations, ideally with scalability guarantees. Performance is improved with top-of-the line research on fast data management algorithms; their efficiency, however, is contingent on seamless collaboration between the database software and hardware and storage devices...Nowadays performance is synonymous to scalability; and scalability, in turn, translates

> into sustainable and predictable use of hardware resources in the face of embarrassing parallelism and deep storage hierarchies while minimizing energy needs—a multi-dimensionally challenging goal.

In the same conference, Juan Loaiza, Senior Vice President of Systems Technology, Oracle, explained how engineering database hardware and software together has been achieved in terms of performance, reliability, and security for the database and rest of the modern data processing stack. The above facts of informatics were kept in mind while hardwares were procured and made functional for SECC.

Central setup hardware architecture, central setup hardware platform, central setup hardware sizing, network sizing and server sizing were kept in view at the time of hardware procurement and configurations. The hardware architecture was designed to work at three layers. The front end servers to host the application for interface with the users (DCCs/District CCs). During the data collection phase, multiple servers will be used for data upload and monitoring applications. Once the data capture exercise was over, some of these servers will be used for hosting the MIS application, while others will be used for bulk report generation. The middle layer servers were used for importing of the uploaded data files into MoRD/MoHUPA database. Once the data collection exercise is over, these servers are to be used for business process implementation like identification of beneficiaries, generation and export data etc. The backend servers will host the database. These were configured for speedy and reliable availability. As this is a time bound & mission critical application, so online SAN replication was setup between NIC NDC Shastri Park & NIC Pune data centres. However as a data centre policy for safeguarding the data against any loss, the backup of the data was regularly taken at the DC site. At central setup hardware platform, it was decided to use open source architecture, so Linux-advance server software was used and these applications were built using open source J2EE framework with JBOSS as the application server.

Enterprise database: SECC is using Massively Parallel Processing (MPP) open source database to manage huge database size of 124 crore records in 24.80 lakh EBs. Additionally, PostGreSQL is used as the database server for transaction requirements.

Central setup hardware sizing resulted in arriving at data size of 6 MB or 48 Mb approx for an EB. SAN for entire solution was done of 250 TB (unformatted in RAID 6 each at data centre as well as at disaster centre) for both rural and urban database. Lastly, server sizing for around 10,000 charge centres, 640 Districts and 35 States/UTs leads to requirement of at least 4 to 6 front end/database servers. For meeting the needs of all types of computing requirements including database management and MIS, the following hardware are installed at NIC National Data Center (NDC) at Shastri Park, New Delhi.

(i) Six numbers of 4 Proc- Blade Server-II (Full Height) x86 Processor based motherboard with four 8 Core processors 6172 2.1 Ghz speed, 64 Bit, 12MB L3 cache, 96 GB DDR-III RAM memory with ECC, 2 x 300 GB SAS HDD, Quad Gigabit Ethernet with WOL in a redundant configuration.

(ii) Four numbers of 4 Processor-Server Blade-II (Full Height) x86 Processor based motherboard with four 8 Core Proc 6172 2.1 Ghz speed, 64 Bit, 12MB L3 cache, 128 GB DDR-III RAM with ECC, 2 x 300 GB SAS HDD, Quad Gigabit Ethernet with WOL in a redundant configuration.

(iii) Two numbers of Intel 4 Processor-Server Blade-II (Full Height) x86 Processor based motherboard with four 8-core processors 6172 2.1 Ghz speed, 64 Bit, 12MB L3 cache, 128 GB DDR-III RAM with ECC, 2 x 300 GB SAS HDD, Quad Gigabit Ethernet with WOL in a redundant configuration.

As part of the NIC Data Centre policy, the website should have DC setup also. Six high end blade servers were placed at NIC Pune Data Centre in DR with following configurations:

Two numbers of Intel 4 Processor-Server Blade-II (Full Height) x86 Processor based motherboard with four 8-core processors 6172 2.1 Ghz speed, 64 Bit, 12MB L3 cache, 128 GB DDR-III RAM with ECC, 2 x 300 GB SAS HDD, Quad Gigabit Ethernet with WOL in a redundant configuration.

Summary and Conclusion

The core of the SECC is the demographic and socioeconomic data of the residents. The collection process also generated a large amount of metadata and logs. The large amount of data collected and metadata generated required layered approach of data management. A single database management system could not suffice and a combination of database management systems were used at various layers for getting the desired performance.

The reporting requirements from the data were typically categorised into—Metadata reports for getting the status and quality of the process, data quality reports for getting a sense of quality of data collected. Multidimensional household level reports for getting the deprivation and vulnerability at the household level and individual level reports for getting the demographic details and individual's information. The access to the reports was restricted through role based access control system with hierarchical authorisation process. Tools were developed and made available for all the stakeholders including the field agencies to online monitor the daily progress, pendency, dependency and data quality.

The database systems needed recalibration a few times to take care of the additional field requirements like introduction of verification/ correction process, master data reconciliation etc. The team had a lot of learnings on designing and managing such a large database for the entire population involving daily transactions from more than 10,000 charge centres.

Annexure A-11.1

List of Monitoring Reports Generated from the SECC MIS with Briefs

1. Status of Field Work in Enumeration Block (EB)/Sub-Block;

a) Enumeration (E)-Stage Summary of households and persons enumerated for completed EB;

b) *Supervisor survey report-1, 2 and 3*: Enumerators and supervisors played the most important role in the whole SECC process. They were the starting point on field data collections. Because they do the field data collection with the help of TPCs, it becomes easy to monitor their activities. Reports on Enumeration (E) Stage and Supervisor survey as mentioned above give details on these activities. For example, at sub-block level, the number of new household entered, new persons enumerated/supervised, household/persons deleted, reasons for household not enumerated, household not enumerated and reasons not given, etc. Such reports proved to be very useful to improve the coverage as well as quality of enumeration and supervision.

c) *SECC coverage of households and population*: Status of SECC coverage in the final stage can be known in both rural and urban areas at EB level. After enumeration, supervision, verification and claims & objection stages, one can know the households added and deleted, total households, numbers of persons added and deleted and total persons.

d) Reports on status of verification and correction of data collection deficiencies, readiness of an EB for draft list publication, status of claims and objections received and disposed, readiness of the EB for data finalisation and final list publication etc. Verification stage was introduced after supervision stage to address ground truthing of few critical household/individual parameters including caste. This is done in selected areas where there was prevalence of deficiencies in data collection in this regard. To check that all the data capture process has been completed before the publication of the draft files readiness report was facilitated. Status of publication of draft lists was also made available. During the draft process, the claims and objections received are uploaded to the central system and the status of their disposal is made available.

e) Status of merging of SECC data with NPR data and the local language translation of data of new members enumerated in SECC. SECC database was built over the NPR database. Thus a comparative statement after merger of SECC and NPR through many incites, including coverage. During field enumeration with TPCs, the data for new members was collected in English language. Subsequently the data was converted to local language by CPSUs at central facilities using the local language data entry process.

2. Data Quality Reports

This category of reports included any duplicates identified in data, jurisdiction related discrepancies, caste correction quality monitoring etc.

- AHL_TIN is the numbering system for households and individuals adopted for SECC. The household/individual identification number is allotted based on the unique location codes in the location directory and unique SECC household number. In order to avoid duplicates in AHL TIN and its reflection on SECC, a report is generated for ready reference.
- As data is captured through electronic media, possibilities of manipulations can't be avoided. However, the SECC MIS was alert by running an algorithm to detect such cases and intimated it to appropriate authorities.
- Location of an EB may be in rural or urban area and this jurisdiction changes dynamically. In the master directory, each EB is identified either as rural or Urban. Since the SECC questionnaire of rural and urban are different, cases were observed where wrong questionnaire (rural versus urban) were canvassed in few EBs. The same was reported online through these reports.

3. Basic MIS

The purpose of the basic MIS was to provide real time information on certain comparable and verifiable attributes of a particular location. These reports were used by the field authorities to measure the correctness of the information gathered and uploaded for the specific area. These reports included distribution of the population according to gender, caste group (SC/ST/Other), PTG (primitive tribal group) population, bonded labourer population, MS (manual scavenger) population, distribution of the households according to type (institutional household, houseless, with shelter etc.), categorisation of the EB (slum, non-slum, uninhabited etc.), comparison of SECC households and population with NPR households and population etc.

- *Summary of new members enumerated*: The enumerator and DEO (data entry operator) visits each EB with TPCs that were preloaded with NPR images that contained demographic details of the respective households. As there was time lapse between NPR operation and SECC, there was need for updating the demographic details, including entering details of new members. The new members may be married females joining the household, new born or/and any other types. This report gives information on these in both rural and urban areas.
- *Summary of household type for completed EB*: Houseless (with shelter) and houseless (without shelter) are main targets for assistance for both MoRD amd MoHUPA. Thus the households reported in these categories at village and town level are listed in this report for close monitoring and validations.
- PTG, BL and MS reports: PTG, BL and MS are crucial household categories qualified for categorisation as 'included' beneficiaries in rural area for schemes of Ministry of Rural Development. Thus enumeration of these categories was closely monitored during SECC for coverage as well as correctness.
- *Summary of SECC and National Population Register (NPR) differences*: Though enumeration was done at different point of time, this report was generated by comparing the NPR database with SECC database after all field data of the latter was uploaded. EB/Sub-Blocks having less or more coverage of numbers of persons under SECC were identified and the district and DCC authorities were alerted to find out the reasons for less coverage and take corrective measures.
- *Slum and non-slum EB*: Slums are urban phenomenon but may exist in rural areas of urban peripheries. Similarly are classified as Type 1, Type 2 and Type 3. In order to ensure the coverage of all the inhabitants in these slums, this report has been generated.
- *Uninhabited EBs certified*: Due to various reasons, at the time of house listing for NPR, an EB may have identified as inhabited. But during SECC field survey, the enumerator may report it as uninhabited EB. This may be deliberately to reduce workload or actually uninhabited. Thus to capture the reality, a certificate from State government is required to declared an EB as uninhabited. This report helps to identify the uninhabited EBs and the status of certifications of them.
- *For-Read implementation and Master concordance report*: House listing blocks of NPR have undergone changes due to bifurcations or merging at the time of SECC. FOR-READ statement are provided by ORGI in case of such EBs due to change in their unique location codes. The data of FOR EBs are migrated to READ EBs through a process. Numbers of such EBs successfully converted and that could not be converted are found in this report.

N.K. SAHU

Summary, Conclusions and Lessons

Socio-economic and Caste Census (SECC) was a huge and complex project. Its social and political implications made it unique and challenging. The institutional arrangements to bring a number of stakeholder agencies within a federal democratic ecosystem derived considerable value for the project. Many technological challenges were faced due to introduction of electronic device called TPCs (tablet personal computers) and the web-based MIS (management information system). How this project was organised and the intended outcomes derived was explained in previous chapters. During the implementation of this project, every stakeholder, both internal or external and core or peripherals, derived many lessons. In this last chapter, along with summary conclusions, our objective is to list out a few lessons that are not only important, but also helps us in providing valuable inputs to organise similar projects in future.

Secrecy and Privacy of Data

One of the most important provisions of the Census Act 1948 is that it makes provisions for the maintenance of secrecy of the information collected in the census of each individual. The Act requires strict secrecy to be maintained about the individual's record which should not be used for any purpose against the individual except for an offence in connection with the census itself. The census records are not open to inspection and also not admissible as evidence. The answers ascertained at the census can be used only for statistical purposes in which the individual data get submerged and aggregated.[1] On the other hand, when the objective is to use the

1. *http://www.gktoday.in/blog/salient-features-of-census-act-1948/#Secrecy_of_Information*

individual or household level data, like in SECC, for identification of target beneficiaries, public scrutiny became imperative in a democratic governance system. This does not mean that the privacy of individuals is compromised. The SECC processes ensured that privacy concerns were taken care by making it clear to the surveyed individuals and households at the start of enumeration and in the acknowledgement slip issued after completion of canvassing.[2] Even after recording the consent or no-consent of respondents, a second chance was given to everybody through advertisements in local newspapers to express their desire not to share the household information/data collected during SECC enumeration at the time of publication of the draft list and thereby in the public domain. Though to a great extent the privacy requirements on SECC data were addressed, the real test would be faced when the privacy and secrecy aspects are put to scrutiny by the legal systems/institutions of the country. Secondly, at the time of utilising the data for identification of targeted beneficiaries, those who concern about privacy and secrecy issues, may raise alarms.

Stakeholders' Role Optimality

Throughout the book including in chapter 3, we have noticed how multiple of stakeholders deliver various project inputs/services and adopted a communication and engagement mechanism to derive to the desired outcomes. When the main stakeholders are very powerful, though they work within the same management/ administrative framework, the project suffers in terms of achieving efficiency. Inordinate time over-run resulted as the needs of the main stakeholders—Ministry of Rural Development (MoRD), Ministry of Housing and Urban Poverty Alleviation (MoHUPA), Office of Registrar General, India (ORGI) and State governments, not only differed, but also the resource in their hands and flexibilities at their disposal. These stakeholders can be classified broadly in to two groups; ORGI as a standalone stakeholder and the remaining as a

2. In the SECC acknowledgement slip, the respondent certifies that: "I am aware that my personal/ household level data, except religion and Caste/Tribe will be published. I give my consent/I do not give my consent for the same (Strike off whichever is not applicable)".

group of stakeholders. ORGI with the objective to conduct the caste census that is a continuum of the decennial population census was concerned to match these two operations conducted at two point of time with two sets of service providers and field operators on the one hand and to maintain secrecy and privacy under the Census Act 1948. The other group of stakeholders (MoRD, MoHUPA and State governments) with objective to conduct a census operation of each households can be utilised to identify targeted beneficiaries under their respective schemes and government interventions. Though this group of stakeholders have vast experience (except MoHUPA) in organising such census operations in past, they don't intend to rely on any legacy data or methodology like that of the ORGI. Their slate was clean. However, for them, addressing privacy and secrecy concerns of every respondent on the one hand and putting the data to public scrutiny on the other hand, was challenging. Thus when all these stakeholders agreed to work together, though the conflicting objectives were not compromised, the outcomes were delayed considerably, and probably, the efficiency in operation was compromised to some extent.

TPC: The Manufacturing Issues

Experimenting with new ideas of governance was done world over including in India. However, such experimentations are done mostly in discrete manner with projects/schemes/interventions etc. in government sector. Most of such projects are local in nature having low financial implications. Such experiments were replicated or expanded based on evidence based evaluations. This was not the case with the decision to go for a census operation, that's too in a vast and complex country like India, where the data collection, storing and transmission is to be done on TPCs. A reading of chapter 5 have highlighted in subdued tone the difficulties in designing, components sourcing, manufacturing and quality testing an electronic devise that was in limited operation in the market. Though the government assigned this task to a reliable public sector electrical and electronic good manufacturing giant, non-existence of standard TPCs models and appropriate manufacturing facilities, leads to complications in the early stages of the SECC field operations. Rectifications and improvements were carried out, but with lapse of considerable time. The cost-plus pricing adopted for the TPCs by a monopoly public sector that was supplied to its owner-customer may not reflect the true input-output technological mixes in achieving efficiency. Such a situation could have probably avoided in a economy with multiple of suppliers of standard products (TPCs). But, as using TPCs in SECC is an experiment, the product came with its problems.

TPC: The Vulnerable Electronic Devise

In the 1st week of June 2016, news came that the Facebook CEO Mark Zuckerberg's Twitter account was hacked. 171 million users of the giant Russian social media service VK was hacked. Similarly the Google CEO Sundar Pichai's Quora account has been hacked by the same group (OurMine Team) in June 2016. Few months back, WhatsApp, a mobile based messaging application, updated its messaging app so that every text message and voice call were encrypted end-to-end for the company's one billion users. Another type of data/information hacking or stealing and dissemination was demonstrated by Julian Assange by establishing Wikileaks, a web site that posts secrete but sensitive materials having international ramifications. Edward Joseph Snowden an American computer professional, former Central Intelligence Agency (CIA) employee, and former contractor for the United States government copied and leaked classified information from the National Security Agency (NSA) in 2013 without prior authorisation. His disclosures revealed numerous global surveillance programmes, many run by the NSA and the Five Eyes Intelligence Alliance with the cooperation of telecommunication companies and European governments. There are many such instances in the world of information technology. Hacking, manipulating, cajoling and many such activities are possible and it would be possible as it is the very nature of the electronic ecosystem that offers the scope of such engagements.

During the use of TPCs in the SECC project for recording the household survey responses, few instances of data manipulations were detected. There were two types of manipulations, firstly, the enumerator/supervisor and DEO (data entry operator) have not really visited the respondents' house, but recorded the reply. Avoiding such manipulating is easy as the recording device is an electronic store. Inbuilt geo-tagging would indicate whether the enumerators/supervisor/DEO has reached the distinct and designated location (the premises of the respondent). This geo-tagging usually consists of latitude, longitude coordinates and time references. If the enumerator/supervisor and DEO visits a location (village, town, Enumeration Block, etc.), the latitude and longitude coordinates would be captured automatically in the TPCs along with the time and duration of canvassing the questionnaire. This would authenticate the visit to the designated location by the enumerator/supervisor and DEO. Secondly, even though the enumerator/supervisor and DEO actually visit the location, they may take only a few minutes to load the data (standard reply may be imported from a storing devise like pen drive). To neutralise such manipulations, an average minimum time limit may be set in-built in the TPCs. This would ensure the visit and duration of survey of each respondent. Some more conditionality can also be incorporated in the software of the TPCs to ensure authentic feedback from the respondents. The second type of manipulations that is possible relate to posting dummy data without doing real entries. This is probably the biggest challenge in such an exercise. As human interventions can be manipulated in the operation of any electronic devise, countering such activities is also possible by smart and alert human interventions. Without going in to the details, in SECC project also, the manipulative human interventions were kept in control through counter human responses as and when such malpractices were detected. However, a smart human counter response mechanism would have been strategised at planning stage of such a project to offset the detrimental moves of some ill-intended human/agencies.

Human at Work: Variance in Capabilities and Questionable Sincerity

Tens of thousands of humans worked for SECC to perform various duties. At the ground level, the lowest level worker was the enumerators and DEOs and above them were supervisors and verifiers. The combine strength of this human resource runs to above one million. Above them, numbering few thousands are the charge officers, district/sub-district nodal officers/agencies, state nodal officers/agencies and national nodal officers/agencies. Most of these human resources were trained and put to various specific assignments as required for SECC project. All of them were expected to deliver specific outputs and services within a framework of roles and responsibilities. But at the execution stage, due to variance in capabilities their performance differed. Four states/UTs reached the last lap in 2012, 10 in 2013, 19 in 2014 and last 3 in 2015.[3] Though one of the main reasons in delay in completion of the project is variance in capabilities, one cannot deny the questionable nature of their sincerity. Despite two stages (enumeration and supervision) being implemented at ground level and a data-view module was made available to monitor the quality of household level data at charge centres that were opened at each CBO (community block office) or *Tehsil* Office; many crucial parameters were recorded much contrary to the quality expectations. And because of this an additional verification module was introduced and executed. Thus ensuring sincerity of those who were working for this project was problematic. As there is a mix of permanent public servants and temporary contractual workers in this project, ensuring quality outcomes was difficult. At times, tacit coalitions between these two actors were marked and corrective actions were taken. However, for a huge and complex project that is highly-human skilled intensive on the one hand and technology intensive on the other, proper designing of the contractual agreements and enforcing them properly a sine-quo-non for its success.

Federal Governance: Each Level has its Peculiarities

Being a federal democratic country, India faces many problems while implementing its schemes and projects. Many of the schemes and projects are prioritised, designed and sponsored fully or partially

3. *http://rural.nic.in/netrural/rural/sites/downloads/BPL_Census/SECC_status_Draft_List_Printing_English.pdf*

by union government with sharing of funding with states. Even states have their own schemes and projects as per their priorities. But targeting individual beneficiaries in general under all these schemes is common, India being a welfare state. However the methods of selection of targeted beneficiaries may differ between central schemes and state schemes. While union government would like the selection of beneficiaries to be done as per the methods of their choice, the states may like to adopt their own criteria. These conflicting interests were observed while executing the SECC project. As this project was fully funded by the union government, every state was expected to work wholeheartedly for completion of this project. However, local conditions prompted few states to insist for local solutions. As far as possible, such demands were met. The main issues relates to adopting local/state languages while canvassing the schedules and publishing these for public scrutiny. Secondly, sub-dividing the administrative units from state to village and/or town level demanded customised treatment. Even though such requirements were met to a large extent, few states were not satisfied as they were not free to modify either the survey methods or the software and hardware configurations.